www.wadsworth.com

www.wadsworth.com is the World Wide Web site for
Thomson Wadsworth and is your direct source to
dozens of online resources.

At *www.wadsworth.com* you can find out about
supplements, demonstration software, and student
resources. You can also send email to many of our
authors and preview new publications and exciting
new technologies.

www.wadsworth.com
Changing the way the world learns®

The Heart of Counseling

A Guide to Developing

Therapeutic Relationships

JEFF L. COCHRAN
State University of New York College at Brockport

NANCY H. COCHRAN
Blossom Road Psychotherapy Associates

THOMSON
™
WADSWORTH

Australia • Canada • Mexico • Singapore • Spain
United Kingdom • United States

THOMSON
WADSWORTH

The Heart of Counseling: A Guide to Developing Therapeutic Relationships
Jeff L. Cochran and Nancy H. Cochran

Executive Editor: *Lisa Gebo*
Assistant Editor: *Alma Dea Michelena*
Editorial Assistant: *Sheila Walsh*
Technology Project Manager: *Barry Connolly*
Marketing Manager: *Caroline Concilla*
Marketing Assistant: *Rebecca Weisman*
Marketing Communications Manager: *Tami Strang*
Signing Representative: *Deborah Van Patten*
Project Manager, Editorial Production: *Catherine Morris*

Art Director: *Vernon Boes*
Print Buyer: *Lisa Claudeanos*
Permissions Editor: *Joohee Lee*
Production Service: *Sara Dovre Wudali, Buuji, Inc.*
Compositor: *Cadmus*
Copy Editor: *Kristina Rose McComas*
Cover Designer: *Paula Goldstein*
Cover Image: *Digital Vision/Getty Images*
Text and Cover Printer: *Transcontinental Printing/Louiseville*

For more information about our products, contact us at:
Thomson Learning Academic Resource Center
1-800-423-0563

For permission to use material from this text or product, submit a request online at
http://www.thomsonrights.com

Any additional questions about permissions can be submitted by email to
thomsonrights@thomson.com

Library of Congress Control Number:
2005922782

ISBN 0-534-62577-0

Thomson Higher Education
10 Davis Drive
Belmont, CA 94002-3098
USA

Asia (including India)
Thomson Learning
5 Shenton Way
#01-01 UIC Building
Singapore 068808

Australia/New Zealand
Thomson Learning Australia
102 Dodds Street
Southbank, Victoria 3006
Australia

Canada
Thomson Nelson
1120 Birchmount Road
Toronto, Ontario M1K 5G4
Canada

UK/Europe/Middle East/Africa
Thomson Learning
High Holborn House
50/51 Bedford Row
London WC1R 4LR
United Kingdom

Latin America
Thomson Learning
Seneca, 53
Colonia Polanco
11560 Mexico
D.F. Mexico

Spain (including Portugal)
Thomson Paraninfo
Calle Magallanes, 25
28015 Madrid, Spain

This book is dedicated to the many clients and students that we have come to know through deep sharing of feelings and ideas. Without this experience, this written work would not exist. Thank you all.

Brief Contents

Preface   xv

Introduction   1

1   Eleven Concepts—Roots That Ground and Grow with
    The Heart of Counseling   5
2   The Rich and Subtle Skills of Therapeutic Listening   19
3   Striving for Empathy   43
4   Expressing Empathy   60
5   Striving for and Communicating Unconditional Positive Regard   78
6   The Delicate Balance of Providing Empathy and UPR in a
    Genuine Manner   102
7   Logistics of Getting Started with New Clients   124
8   Initial and Ongoing Structuring of Therapeutic Relationships   150
9   When Clients Need Help Getting Started   167
10 Managing Client Crises with Therapeutic Relationship Skills   183
11 Ending Therapeutic Relationships   221
12 Therapeutic Relationships across Cultures   250
13 Growing Your Therapeutic Relationship Skills and the Core
    Conditions of Counseling   266

References   280

Special Reference List for Pre-Chapter Quotes   285

Skill Support Resource A: Brief Summary Notes for Chapter 1: Eleven
Concepts—Roots That Ground and Grow with *The Heart of Counseling*   287

Skill Support Resource B: Do's and Don'ts of Reflective Listening
and Expressing Empathy   289

Skill Support Resource C: Sample Initial Session Report Items   291

Skill Support Resource D: Essay Assignment for Particular Use in Readying
Your Self-Awareness for Counseling across Cultural Differences   292

Index   295

Contents

PREFACE xv

INTRODUCTION 1

Why the "Heart" of Counseling   1

A Few Notes about Us   2

A Note About Our Theoretical Base and Background   3

Important Notes on Case Examples   3

How to Use This Book   4

1 Eleven Concepts—Roots That Ground and Grow with *The Heart of Counseling*   5

Primary Skill Objectives   5

Focus Activity   5

Introduction   6

Important Guidance for Your Study of These Concepts   6

   Avoid Intellectual Overload   6

   Remember That Experience Is the Best Teacher and Communicator   6

Eleven Underlying Concepts   7

   Self-Actualization   7

   When Blocks Come into One's Path toward Self-Actualization and Ideal Maturity   9

   The Capacity for Awareness, to Reason, Question, and Choose   9

   Interpretation of Experience and Development of Self-Concept   10

   Awareness of Existence, Choice, and Questions of Self-Worth   12

   Self-Responsibility is Anxiety-Provoking   13

   Awareness of Aloneness   13

Emotions Are Useful and Necessary for Growth   14

Every Action Is a Choice of Destiny   15

The Internal World   16

Locus of Control and Evaluation, and Being   16

Activities and Resources for Further Study   17

2   The Rich and Subtle Skills of Therapeutic Listening   19

Primary Skill Objectives   19

Focus Activity   20

Tuning in and Listening   21

The Many Levels and Nuances of Reflecting   22

The Broad Skill of Reflection: From Paraphrasing to Themes to Confrontation, Challenge, Summary, and Beyond   22

The Broad View of Communication from Clients   24

The Do's and Don'ts of Listening Therapeutically   25

Overview of What You Are Communicating with Your Actions   25

Do's   25

Don'ts   26

Important Explanations of the Behaviors of This Do's and Don'ts List   26

Your Listening Body Language   26

Reflect Your Perception of What Your Client Communicates   27

Reflections as Declarative Statements   28

The Question of Questions or Questioning Tone   29

Keep Your Reflections Short, Whenever Possible   32

The Special Issue of Verbatim Reflections   32

Focus Your Reflections on Your Client's Main Point, or the Things Communicated That Seemed Most Important, Most Emotionally Laden for Your Client   33

Be Prepared for and Accept Corrections   34

The Issue of Interruptions   34

Allow Your Clients to Own Most Silences   35

How Therapeutic Listening Differs from Listening Outside of Counseling and from Nonlistening   36

How and Why Listening Therapeutically Works   36

What's Difficult about Listening—Common Errors and Problems
in Listening for Beginning Counselors   38

    Common Interfering Thoughts   38

    Therapeutic Listening Is Not Normal and Can Feel Odd   39

    The Urge to Fix Immediately   39

    Sometimes Clients Want a Quick Fix   40

    *The Desire to Ask Informational Questions That Are of
Interest to You*   40

    Stories   40

    Returning to Important Communications   41

Multitasking with Therapeutic Listening   41

Activities and Resources for Further Study   41

3 Striving for Empathy   43

Primary Skill Objectives   43

Focus Activity   44

Introduction   45

What Empathy Is and Isn't   46

    Understanding What Empathy Is by Considering What It Is Not   46

    Intricacies of Empathy   47

A Sample of the Preponderance of Literature Supporting
and Clarifying the Importance and Power of Empathy   48

Why Empathy Is Important, Powerful   50

    Connecting at the Core   50

    Joining on a Scary Journey   50

    Self-Awareness   51

    Self-Experience   51

    A Profoundly Different Relationship   52

    Joy in Connecting   52

    Furthering Communication and Connection   53

What Gets in the Way of Empathy   54

    Habit   54

    Fear of Feelings   54

    Misattributed Responsibility   55

Letting Go of Control   56

Activities and Resources for Further Study   57

4 Expressing Empathy   60

Primary Skill Objectives   60

Focus Activity   61

Various Ways to Express Empathy   62

Matching Client's Tone   62

Facial Expression and Body Language   62

The Most Overt Means—Words   63

Do's and Don't of Expressing Empathy   64

Overview   64

Do's   64

Don'ts   65

Explanations and Discussion from This Do's and Don'ts List   65

Focus Your Attention Primarily on Client Emotions, Secondarily on Thoughts and Actions   65

Strive to Feel with Your Client, to Feel What Your Client Feels   65

Reflect Client Feelings and Underlying Thoughts through the Tone, Facial Expressions, Body Language, and Gestures That Come Naturally to You   66

Reflect Client Feelings with Words for the Emotions You Feel with Them, When Natural   66

Reflect Feelings and Underlying Thoughts That You Perceive Your Client to Imply   66

State Your Empathy in Declarative Statements and, When Unsure, State Your Empathy with More Tentative Declarations   67

Use Reflections to Restate Client Feelings and Underlying Thoughts More Clearly, Directly, and More Precisely and Concisely   67

Be Prepared for and Accept Corrections   68

Don't Let Your Words for What You Feel with Clients Come Out Sounding Like Assessments   68

Don't Respond from a Hidden Agenda of What You Believe Clients Should Realize   69

Don't Do Most of the Talking   69

Don't Make "Me Too" or "Must Feel" Statements   69

Nuances of Expressing Empathy That Don't Quite Fit
Our Do's and Don'ts List　70

　　*Remaining Animated, Natural, and Spontaneous
　　in Expressing Empathy*　70

　　Variation of Tone in Expressing Empathy　70

　　Responding to Implied Emotions　70

　　Responding to Unpleasant Emotions　71

　　*Discerning When to Respond More to Emotions and When
　　to Respond More to Content*　72

Common Difficulties, Pitfalls, and Dead Ends　73

　　Thinking of the Word, Rather Than Feeling with Your Client　73

　　Trying Too Hard to Get It Just Right　73

　　A Limited Vocabulary for Feelings　74

　　The Problem with Claiming Understanding or Shared Experience　74

　　Personal Confidence and Faith in the Counseling Process　75

　　A Lack of Unconditional Positive Regard　75

Activities and Resources for Further Study　76

**5 Striving for and Communicating Unconditional
Positive Regard　78**

Primary Skill Objectives　78

Focus Activity 1　79

Focus Activity 2　79

Introduction　80

　　*Paths toward Holding Others in Reasonably Consistent
　　Positive Regard*　80

　　The Tandem: Empathy and Unconditional Positive Regard　81

What UPR Is and Isn't　82

　　Beginning Thoughts on What Unconditional Positive Regard Is　82

　　What UPR Is Not　84

A Sample of the Literature Supporting and Clarifying the
Importance and Power of UPR　86

Why UPR is Important, Powerful　86

　　Self-Acceptance = Change　87

　　UPR = Full Expression of Emotions　88

As We Accept Our Clients, Our Clients Come to Accept Themselves 88

A Safe Environment 88

Evaluation by Others Can Be a Poor Guide for One's Self 89

Rewards for the Counselor and Client 89

How UPR Is and Is Not Communicated 90

What Gets in the Way of Having and Communicating UPR 92

Having an Agenda for Your Client 92

Counselors Believing They Know Better Than Their Clients 92

Burnout 93

Lack of Self-Acceptance 95

Counselors Inadvertently Seeking to Fill Their Own Needs through Clients 95

The Analytic Mind 95

Some Clients are Hard to Like 96

Clients Doing or Saying Things That Run Counter to Their Counselor's Moral Constructs 96

Initial Judging Thoughts 97

How UPR in Counseling May Relate to UPR outside of Counseling 97

Activities and Resources for Further Study 98

6 The Delicate Balance of Providing Empathy and UPR in a Genuine Manner 102

Primary Skill Objectives 102

Focus Activity 103

Introduction 103

What Genuineness Means and Does Not Mean 104

A Sample of Literature Supporting and Clarifying Genuineness and Its Importance 105

The Importance of Genuineness in Counseling 107

Keeping Therapeutic Listening, Empathy, and UPR Real: A Therapeutic Relationship with a Real Person 107

The Connection to and Role of Genuineness in the Set of Core Conditions 107

Modeling   108

Creating a Safe Place for Emotional Honesty   109

How Counselor Genuineness Is and Is Not Communicated   110

Declarations of Genuineness Are Rarely Helpful   110

*Sometimes Counselors' Experiences of Clients "Bubble up"
or Cannot Be Hidden*   110

*State Your Reaction When Your Reaction to Clients Interferes
with Your Empathy and UPR*   111

Being Who You Are in the Phrasing of Your Reflections   113

*Make Only Careful, Judicious Self-Expressions, Beyond Your
Ever-Present Empathy and UPR*   114

What Makes the Delicate Balance of Providing Empathy and UPR
in a Genuine Manner Difficult   114

The Errant Thought—I Am Who I Am   114

The Challenge of Clients Who Are Hard to Like   115

*The Need for a High Level of Self-Development
for This Counseling Skill*   115

*The Question of Expressing Your Positive and Negative
Experiences of Clients*   117

*The Need for High-Level Observational Skills, Therapeutic
Listening, and Empathy*   118

Clients Who Ask for Your Experience of Them   119

*The Need to Balance Freedom That Optimizes Personal Connections
and Allows Experiences to Bubble Up into Expression with Avoiding
Influence That Limits Client Expression*   121

A Closing Thought on Genuineness   122

Activities and Resources for Further Study   122

**7 Logistics of Getting Started with
New Clients   124**

Primary Skill Objectives   124

Focus Activity   125

Introduction   125

Gathering Information and Understanding for an Initial Session
Report Using Your Skills in Therapeutic Listening, Empathy, UPR,
and Genuineness   126

Learning about Initial Counseling Sessions from Situations That Require Intake or Initial Session Reports   126

Practical Reasons for Information Gathering in Initial Sessions   127

Incorporating Counseling-Related Assessment with Your Skills in Therapeutic Listening, Empathy, UPR, and Genuineness   129

The Issue of Writing Notes during Sessions   130

Writing an Initial Session Report   130

Identifying Information   131

Presenting Problem/Concerns   131

History of the Problem/Previous Interventions   131

Reason for Coming to Counseling Now   132

Alcohol/Drug Use and/or Medical Concerns   133

Related Family History/Information   133

Major Areas of Stress   134

Academic/Work Functioning   134

Social Resources   134

Initial Impressions or Understanding of the Person and Concerns   135

Treatment Plans   137

Additional Notes on Thinking through Initial Impressions or Understandings of the Person and His or Her Concerns and Treatment Plans   139

Keeping Ongoing Case Notes   140

Common Dilemmas or Situations in Getting Started with New Clients   141

A Need to Know What to Expect   141

Anxiety   141

An Explanation(s) of Counseling That Helps Clients Begin   142

Information Your Clients Should Know When Getting Started   142

Confidentiality   142

Who and/or Why Referred   143

Potentially Helpful Information Related to the Presenting Problem   144

Goals   144

Problems with Goals   144

The Thinking behind Our Solutions to Establishing Goals   145

Reasonable Goals   146

Unreasonable Goals   147

Communicating Goals   148

Activities and Resources for Further Study   148

8 Initial and Ongoing Structuring of Therapeutic Relationships   150

Primary Skill Objectives   150

Focus Activity   151

Introduction   151

Logistics   151

Session Length and Ending Sessions   151

*Time Warnings and the Importance of Letting Clients
Own Their Endings*   152

Exceptions to Ending on Time   154

*The Awkwardness of Giving Time Warnings
and Guidance for This*   154

*Possibilities of Varying the Time Warning
Structure for Some Clients*   155

A Few More Suggestions on This Time Thing   155

Helping Clients Understand the Structure of Interactions in
Counseling or How Counseling May Work for Them   155

A Client Who Asks for Guidance   156

A Client Who Seems to Insist on a Quick Solution   157

A Client Who Has Great Discomfort with Silence   159

A Client Who Attends Sporadically   160

A Client Who Just Does Not Know Where to Start   161

Higher-Level Reflections Can Also Help Clients Understand How
to Use Counseling   162

Example 1   162

Example 2   163

Explaining Therapeutic Relationships or Use of Counseling to
Significant Persons in Clients' Lives   164

Problems for Beginning Counselors in Explaining Clients' Potential
Use of Counseling   165

Activities and Resources for Further Study   166

9 When Clients Need Help Getting Started   167

Primary Skill Objectives   167

Focus Activities   168

   Activity 1   168

   Activity 2   168

Introduction   168

Mistakes You May Make That Inhibit Your Clients' Beginning Use of Counseling   169

   Trying Too Hard or Worrying about Motivating Clients to Make Rapid Use of Counseling   169

   Not Recognizing That Your Client Began to Use His or Her Therapeutic Relationship with You As Soon As You Began to Provide It   170

   Lack of Acceptance   171

   Pedantic Reflections   172

   Reflections That Sound Like "Aha" Conclusions   172

   Slipping into Questions   173

   When in Doubt, Please Review   173

Examples of Counselor Actions That Help Clients Struggling in Starting   173

   Start Where Your Client Is   173

   Respond to the Level of Emotion Each Client Expresses   174

   Remember the Uniqueness of Each Client's Pace   175

   Remember That It's Natural to Feel Uncomfortable in the Beginning   175

   Give Them Room   176

   Respect Your Client's Pace   176

   See the Big Picture in Your Clients' Communication   176

   Dispel Expectations of a Need for a Problem or Profundity   177

   Sharing Experience in Letting Go   177

   And Finally, Hang in There   177

Asking Questions or Suggesting Topics That Clients May Find Helpful in Their Struggles in Starting   178

   Stating Why a Client Was Referred or Why You Offer Counseling   179

   Suggesting Common Areas of Importance for Discussion   179

*Basing Questions/Suggestions of Information That You
Already Have and Are Interested In* 180

Open Questions 180

Concluding Thoughts on Helping Clients Who Are
Struggling in Starting 180

Activities and Resources for Further Study 181

**10 Managing Client Crises with Therapeutic
Relationship Skills 183**

Primary Skill Objectives 183

Focus Activity 184

    Part 1 184

    Part 2 185

    Part 3 185

Introduction: Therapeutic Relationships as a Source of Power
and Influence to Help Clients Manage Crises 185

Principles of Managing Client Crises with Therapeutic
Relationship Skills 186

    *Self-Responsibility, Dignity, Integrity, and Least-Restrictive
Interventions* 186

    Acceptance 187

    Empathy 187

    Tell Your Client What's Going On with You 187

    Remember to Reflect 188

    Help Your Clients Make Their Plan 189

    Plan Specifically 189

    Err on the Side of Caution 189

    Say the Words 190

    *Respond to the Possible Communication of a Dangerous Situation
As Soon As Possible* 190

Consideration of Assessment Factors 191

    A Plan 191

    Lethality of the Plan 192

    A Means 192

    Preventive Factors 192

    Future Orientation 193

A Sudden Change or Switch   193

Previous Attempts   193

Lowered Inhibitions and Impulse Control   193

Ability to Guarantee Safety   194

And Finally . . .   194

Non-Self-Harm Agreement/Safety Plan   194

A Good Safety Plan Is Time Specific   194

*Relate the Safety Plan to Avoiding Elements within
Your Client's Suicide Plan or Thoughts*   195

Get Rid of the Means   195

Avoid Lowered Inhibitions and Impulse Control   195

Prevent Harm by Contacting Someone Immediately   196

Work Slowly and Carefully   196

And Finally . . .   197

A Case Example of a Client with Mild Suicidal Ideation   197

The Issue of Hospitalization   201

Know Your Local Laws, Guidelines, and Procedures   201

When to Seek Hospitalization   201

Maintaining Maximal Client Self-Responsibility   201

Responding with Empathy to Clarify Intent   201

Counselor Responsibility   202

The Issue of Paying for Hospitalization   202

An Example with a Client Experiencing Strong
Suicidal Ideation   202

An Example with a Client Who Is at Risk of Harming Others   206

Assessment Factors in Determining Risk and Safety Plans
in Domestic Violence Situations, Especially Those That Rise
to the Level of Imminent Danger   212

Physical Violence   212

The Extent of Physical Violence and Any Potential Pattern   212

Triggers and High-Risk Behaviors   213

Children   213

Planning for Safety   213

Common Difficulties for Beginning Counselors Helping Clients
Manage Situations That Are or May Be of Imminent Danger   214

*The Seriousness of the Situation and the Weight
of Decisions—The Danger Itself* 214

"What if I Panic and Know What to Do, But I Forget?" 214

The Pressure of Never Being 100% Sure 215

Discerning the Difference between Your Feelings and Empathy 215

Errors in Empathy 216

Preoccupation with Liability 216

*Having to Let Go and Let Clients Be Responsible—Trusting That
Each Client Will Actually Do the Plan Agreed To* 216

*Self-Confidence and Self-Perception of Competence
to Make Such Decisions* 217

*Coordination with Other Professionals—Fearing Breaking Trust
to Ask for Help* 217

Coordination with Client Loved Ones or Significant Others 218

*The Infinity of Unknown or New Situations for Which There
Is No Script* 218

*Shifting into Crisis Management Panic Mode and Forgetting to
Continue to Build and Use a Therapeutic Relationship
with Each Client* 218

Activities and Resources for Further Study 219

11 Ending Therapeutic Relationships 221

Primary Skill Objectives 221

Focus Activity 222

Introduction 222

The Principle of Independence 223

Planfulness 224

*Committing to Review for Client Readiness to End
throughout Ongoing Work* 225

*Reviewing for Client Progress, Satisfaction, and Decisions
toward Ending* 225

*Setting a Tentative Plan for Counseling, Reviewing Progress,
Satisfaction, and Decisions about Whether and How Long
to Continue with an Initially Reluctant Client* 227

Recognizing the Many Forms of Progress 228

Consideration of Alternative Modes of Planful Endings 232

Counting Down to the Ending 233

Letting Your Clients Know They May Return 233

Telling Your Clients How You See Them in the End or Last Meeting 233

Arbitrary Endings 234

Help Clients Plan for the Premature Ending 236

Counting Down 237

Discussing/Suggesting Continued Work and Progress 237

Special Problems or Situations That Occur with Arbitrary Endings 238

Seeking Feedback in Final Meetings 242

Common Difficulties for Beginning Counselors around Ending 243

Not Wanting to Let Go 243

The Frequent Happy/Sad Endings 243

Seeming to Want Too Much to End 243

Client Reluctance to End 244

Surprise That a Client Seems More Okay with Ending Than You Do 245

Unknown Reasons for Clients Ending and the Temptation for Counselors to Speculate or Blame Themselves for Some Error 245

Great Satisfaction and Joy in Endings 246

Activities and Resources for Further Study 247

12 Therapeutic Relationships across Cultures 250

Primary Skill Objectives 250

Focus Activity 251

Introduction: All Counseling Is Cross-Cultural—But You Have to Reach Out 252

Reaching across Cultural Differences with Your Skills and Your Self 254

Humility 255

Be Wary of Cultural Assumptions 255

Know Yourself through Immersing Yourself 256

Think Broadly 256

The Fairly Foreign World of Children 258

Clients and Others Who May Not See the Value of Counseling 259

Common Problems or Experiences of Beginning Counselors in Counseling across Cultures 261

Opportunity to Experience a Diversity of Clients 261

Difference as Advantage to the Counselor 261

Missing the Feelings for the Cultural Context 261

Missing the Content for the Context 261

Significant Value Differences 262

The Experience of Connecting 262

Sensing a Need for Information and Context Education 262

Reaching Out and Becoming Accessible 263

Activities and Resources for Further Study 264

13 Growing Your Therapeutic Relationship Skills and the Core Conditions of Counseling 266

Primary Skill Objectives 266

Focus Activity 267

Introduction 267

Developing Yourself through Developing Your Skills in Therapeutic Relationships 268

The Question of Multiple or Dual Relationships with Clients 269

Therapeutic Relationship Skills in Consultation 272

With Teachers 272

With Parents and Loved Ones of Clients 273

With Other Professionals from Related Fields 273

Therapeutic Relationship Skills in Job Task Negotiation 274

Therapeutic Relationship Skills in the Classroom 275

Empathy Sandwich 275

Project Special Friend 276

Teaching Clients and Others to Use the Skills of Therapeutic Relationships in Their Relationships 277

"Oh, the Places You'll Go" 278

Activities and Resources for Further Study 278

SKILL SUPPORT RESOURCE A: BRIEF SUMMARY
NOTES FOR CHAPTER 1: ELEVEN CONCEPTS—ROOTS
THAT GROUND AND GROW WITH *THE HEART OF
COUNSELING*   280

SKILL SUPPORT RESOURCE B: DO'S AND DON'TS
OF REFLECTIVE LISTENING AND EXPRESSING
EMPATHY   282

SKILL SUPPORT RESOURCE C: SAMPLE INITIAL
SESSION REPORT ITEMS   284

SKILL SUPPORT RESOURCE D: ESSAY ASSIGNMENT
FOR PARTICULAR USE IN READYING YOUR SELF-
AWARENESS FOR COUNSELING ACROSS CULTURAL
DIFFERENCES   285

REFERENCES   287

SPECIAL REFERENCE LIST FOR PRE-CHAPTER
QUOTES   292

INDEX   295

Preface

As authors, we are instructors and supervisors of counselors, as well as counselors ourselves. We have undertaken to write *The Heart of Counseling* as we have been dissatisfied in seeing students eager to apply various counseling techniques meant to help their clients but attempting this work without a base connection or therapeutic relationship with the person of their client. We find that beginning counselors often face a crisis in personal and therapeutic confidence. Beginning counselors often seem to fear that while they have knowledge of counseling, they won't actually be able to implement it. Partly creating and partly resulting from this crisis in confidence, and confounding the situation, is the contrast between beginning counselors' need for concrete skills versus the fact that counseling concepts, especially those of therapeutic relationships, are highly abstract. One result is that beginning counselors are often tempted to dive straight into counseling techniques aimed at forcing immediate behavior change without first developing a therapeutic relationship that can ensure that clients own the changes they make. When clients do not own their changes, the changes are usually not true and lasting, but illusive and temporary. Additionally, the fact that *The Heart of Counseling* is focused on the skills of therapeutic relationships, instead of techniques aimed at forcing immediate behavior change, clarifies the power and importance of therapeutic relationships in counseling for beginning counselors.

We want this book to help students know that "slow is fast." We wrote it to help students slow down and form deep therapeutic relationships with clients, to find the value in this work to bring about positive change, and to see just how far their therapeutic relationships can take their clients' growth and their own growth.

We learned from play therapist and author, Bill Nordling, "You are the best toy in the playroom" (personal communication). By this he meant that the counselor and therapeutic relationship are what is healing to a child in play therapy, that the toys are merely a medium of communication. Our equivalent of this in adult and adolescents' "talk therapy" is, "*You* are the best tool for your clients." So, we believe strongly in the value of each counselor's self-development. While this book focuses on helping counselors develop concrete, measurable skills, those skills are intertwined and inseparable from counselors' personal development. Therefore, we help readers develop both their selves and their skills, while remaining respectful of each reader's readiness to engage in self-reflection at that point in time.

We see *The Heart of Counseling* as a key learning tool for students of counseling and related mental health fields. We have designed it as a personal learning tool. We write directly to readers and work to anticipate their needs in developing as helpers—partly through observing the needs of our students, ourselves, and our many friends who are mental health practitioners. We address the confidence issue directly by sharing stories of our students' struggles in personal and therapeutic confidence, and in skill application.

It is designed for active learning. Each chapter begins by identifying primary skill objectives, then involves readers in focus activities to orient and help them begin to anticipate and organize what they are about to read. Each section is clear and succinct in order to help readers store and retrieve what they have learned from that section. We provide extensive case examples that are true to our experience, that of our students, and that of our friends in mental health fields. Very importantly to us, each chapter ends with directions for learning activities to help students implement the skills they have begun to learn through reading that chapter. Our overall goal is to take the seemingly abstract and esoteric principles of therapeutic relationships, and make them as clear, concrete, and skill-based as possible. Our goal is not only for students to read about these skills but to become ready to implement them, to become excellent counselors. While this is a text to help counselors form a strong foundation early, it is also a book that counselors can grow with and will refer back to, especially in troubled times, through years of good works.

Acknowledgments

We would like to acknowledge our editor, Lisa Gebo, for her savvy in making this book happen, for her guidance, understanding, and for "getting it." We also greatly appreciate Sheila Walsh and the many others at Brooks/Cole who have helped this book become a reality. We would also like to thank Deborah Van Patten for her support and encouragement of this project.

We would like to thank each of the following reviewers for their time, insight and hard work in reading and critiquing *The Heart of Counseling*: Robert J. Wellman, Fitchburg State College; Geoffrey Yager, University of Cincinnati; Joanne Cohen Hamilton, Kutztown University; Pamela M. Kiser, Elon University; Dana Comstock, Saint Mary's University; John H. Harvey, University of Iowa; Michael M. Morgan, University of Wyoming; and Barbara J. Higgins, SUNY Brockport. We would also like to thank our copy editors, Linda Ireland and Kristina Rose McComas, for their smooth improvements of our words and careful checking of important details and Sara Dovre Wudali for her work in seeing *The Heart of Counseling* through production to print.

We owe a very special thanks to Jo Cohen Hamilton for her conscientious and wise input. She gave suggestions that helped us think through weak areas and clarify blind spots. We also greatly appreciate Barbara Higgins for her input, support, encouragement, and friendship.

We would like to acknowledge our colleagues and friends, Armin Klein, Grace Harlow Chickadonz, Leslie McCulloch, and Muhyi Shakoor, for our conversations, for our time together and connections. We also thank the counselors at Blossom Road Psychotherapy, Vicki Cummings, Geri Stanton, and Craig Bulloch, for providing a beautiful, supportive group practice and playroom where Nancy continues her work with children and families.

We would like to thank Dottie Reed for her many kind assistances.

Finally, but most importantly, we would also like to state our appreciation to our families for their love and support, especially our parents, Joyce Cochran, and Jan and Richard Haldeman.

Introduction

It is only with the heart that we can see,
what is essential is invisible to the eye.

The Little Prince

ANTOINE DE SAINT-EXUPERY

There is an organ in the body that, if it is righteous,
it ensures that the whole system will be righteous. . . .
This organ is the heart.

MUHAMMAD

WHY THE "HEART" OF COUNSELING

The word *heart* in the title of this book is used to convey two main meanings: core and emotion. The word *heart* implies a central core, as in the saying, "Get right to the heart of the matter." Think, for a moment, of the implications of this saying for the counseling profession. We counselors are often trying to figure out how to be more efficient—to "get to the heart of the matter"—while also maintaining a respect and deep caring for each client's individuality. The client's own pace and path to the "heart of the matter" can be easily overlooked, as it is often challenging territory for both the client and the counselor. Thus, the skills we guide you through in this text are presented as the core, or "the heart," of what counselors do, and as the most important and most essential skills in allowing our clients to lead the way. They are the counseling skills that build courage and confidence. All other counselor tasks spring from and revolve around them.

The word *heart* also implies an active investment of emotional energy, as when referring to an athlete who competes with great passion one might say, "She plays the game with heart." Counseling should never lack an investment of the counselor's emotion, or be dispassionate. While you, as counselor, may often become relaxed and more at ease in the work, you must continue to feel and respond to your clients' emotion accurately with compassion and feeling. The best and most efficient counseling is "heartfelt." Responding fully to emotion, both your clients' and your own, is a theme woven throughout this text on the core skills of counseling. Experiencing your clients' emotions with them, while remaining aware of your own as counselor, is the golden road to developing deep caring and making healing connections. Such connections are both challenging and profoundly powerful in our development as counselors.

A FEW NOTES ABOUT US

Jeff is primarily a teacher of counselors at this time. He has worked as a counselor to students at an elementary school, a middle school, and at universities. He has provided counseling and related mental health services at outdoor camps, at a residential treatment facility, and at high schools. Jeff has completed these works in several areas of the United States and overseas. He currently has the honor of supervising and consulting with counselors and mental health service providers in a great many settings. He continues to learn more and more from their work.

Nancy has education, expertise, and experience with a wide range of ages and persons as a school psychologist. She currently works primarily as a counselor in a private practice, serving children by providing child-centered play therapy, and their parents and other loved ones by training in filial therapy and related consultation. She frequently supervises counselors in child-centered play therapy for the National Institute for Relationship Enhancement (NIRE). She is certified through NIRE in child-centered play therapy and child-centered play therapy supervision.

We have been married for 14 years, and in that time we have had the opportunity to write, research, present, teach, work, and play together. Themes of our combined written works have always focused on the importance and power of counseling relationships. We have enjoyed working together with children in schools, as foster parents, and as camp counselors. We have also enjoyed working together with graduate students in the classroom, and on various projects and research ideas. In this work, we have experienced the power of building effective, fulfilling human relationships with others not only as their teachers, but also as their helpers, and their friends. The joy we find together in this work has helped to maintain and improve our own multifaceted relationship. We share our ideas and philosophical musings about life, and the challenges of counseling and the other helping professions quite a bit. Our ideas and beliefs are often complementary, or so similar that we don't know whose idea was whose, or if indeed, the beliefs and ideas we have are spontaneously created between us!

By far, our most important work together is in loving and parenting our wonderful son, Erzhan, whom we adopted during the course of writing this textbook. This is new to us and is joyous and challenging work indeed!

A NOTE ABOUT OUR THEORETICAL BASE AND BACKGROUND

The counseling theories that have been most influential in our education and our work have been the person-centered approach and the cognitive behavioral approach to therapy. Some of the influence that this background has on our text will be evident. We often emphasize persons' self-talk in examples and present counseling as an experience through which persons learn and grow, and through which self-talk and self-perception changes. We emphasize that the counseling relationship is the key factor in promoting meaningful change and growth for the individual client. These foci are not meant to exclude any counseling theory, but are the result of our experience and study as counselors. Regarding theory, we strongly agree with the following quote from Carl Rogers, "There is one *best* school of therapy. It is the school of therapy you develop for yourself based on a continuing critical examination of the effect of your way of being in the relationship" (1987, p. 185).

nice

IMPORTANT NOTES ON CASE EXAMPLES

We provide a great many case examples to illustrate the skills and concepts of counseling. Some case examples were based on clients of ours, some were based on clients of our students, and some case examples are hypothetical or composite examples based on a combination of similar experiences with clients in similar situations. When case examples are based on persons served by one of us, we tell the example using "I" to refer to the counselor. When the case examples are based on persons served by one of our students, or when they are hypothetical, composite examples, we present the example and refer to the helper as "the counselor." All case examples are presented without identifying information or with the information altered significantly to protect anonymity. Still, we'd like to convey that each of these case examples may contain elements of stories from people who continue to remain special to us.

We remember them well, and continue to care about them, respect them, and wish the best for them. Many of the stories, while changed significantly from any specific person's situation, would not be possible without the deep sharing that happened between counselor and counselee. If our presentation of examples seems too brief, it is likely because we attempt to share *just enough* of the stories to demonstrate how you may help others. This textbook would not be possible without the courage and deep sharing of those with whom we have been in counseling relationships.

Additionally, we occasionally use examples of children to help explain something important. Sometimes a child example is simplest and best. However, this book is mostly about counseling adults and adolescents. The same concepts are true in counseling children, but applications should be different. Children's primary mode of communication, especially communication of the depth for counseling, is play not talk. Further, because children are more willing or able to engage in deep empathic connections in counseling than many adults, even more pure therapeutic relationships are possible and compromises in personal responsibility, such as helping child clients get started or explaining what you and they are doing and why, are rarely necessary.

HOW TO USE THIS BOOK

We'd like to challenge you to spend time contemplating the various concepts presented. In most chapters, we suggest a few modes of contemplation, both social and introspective. Idea exchange with others is often very helpful to learning. Thus we give you suggestions for how to implement social contemplation of the concepts and skills we present. Because such group learning is not always possible and because introspective contemplation brings its own values, we also suggest methods of introspective contemplation. For this, we sometimes borrow methods from meditation and creative arts. We hope you will try each and expand into the further contemplation methods that then occur to you for your unique ways of learning.

We encourage you to actively study. Focus yourself on the subject at hand in each chapter with the Primary Skill Objectives and especially the Focus Activity for each chapter. Stop and contemplate each section, rather than just letting your eyes rush over words. Take time to contemplate the big ideas of each chapter, especially those that have struck you most from that chapter. Complete as many of the Activities for Further Study at the end of each chapter as possible, carefully choosing the ones you most need to focus on in your development.

1

Eleven Concepts—Roots That Ground and Grow with *The Heart of Counseling*

> All wisdom is already within us; all love is already within us, all joy.
> Yet, they are hidden within us until the heart opens.
>
> FADIMAN & FRAGER

PRIMARY SKILL OBJECTIVES

- Establish beginning understanding and be able to explain the concepts presented in this chapter in your own words.
- Begin your process of striving to fully understand these concepts and how they fit with the skills of providing therapeutic relationships.
- Develop your understanding of how these concepts fit with your core beliefs at this time.

FOCUS ACTIVITY

Take time to consider, journal, discuss, and/or essay your beliefs about human nature as it relates to counseling. Consider the following questions: What generates behavior, shapes personalities, or creates well-being and mental health? What motivates change? What prevents it? Based on these beliefs, what do you imagine that you may do as a counselor to facilitate positive change and growth in others?

INTRODUCTION

The following concepts underlie the work of therapeutic relationships. They have become increasingly meaningful and essential to us, through our works as counselors and our experiences of life. They connect and provide context for the skills of this text. They are beliefs that bind together all the tasks we hope to accomplish as counselors. They help us understand the skills we employ in our works, and to know how and why those skills are effective. Contemplating and understanding these concepts helps sustain us through difficult moments in our works. We speak personally about these concepts, and from our own experience. As you read through, and contemplate each concept, consider these guidelines, as it is likely that you will react personally and from your own experience to what you read.

IMPORTANT GUIDANCE FOR YOUR STUDY OF THESE CONCEPTS

Avoid Intellectual Overload

We encourage you to read slowly and to contemplate. While these concepts are interrelated, you may not want to read them all at once. This could give you the equivalent of intellectual indigestion. Give yourself time to go slow. Stop to consider each concept and your own views and reactions as you read. Perhaps stop to write your reactions after each concept (e.g., excited agreement with some parts, troubled disagreement with others, questions you'd like answered to help you understand, questions you'd like to discuss).

Remember That Experience Is the Best Teacher and Communicator

There are examples to aid our communication to you throughout this chapter, but many more throughout the skill portions of this book. Additionally, as you practice the skills of *The Heart of Counseling,* the underlying concepts will become clearer.

Your learning and coming to know what you know, and believe what you believe through your practice and work is part of your development as a scholar and counselor. We only introduce these important underlying concepts in this chapter. Your deep understanding, your *knowing,* can develop from practicing and fully considering the skills that are the focus of this book. Fully understanding these concepts may be the ongoing work of a lifetime for you, as it is for us.

ELEVEN UNDERLYING CONCEPTS

Self-Actualization

Every person is on a unique path to optimal existence. Each person's path is unique because each person's set of experiences, and especially each person's interpretation of experiences, is unique. Additionally, if given the opportunity and freedom, people will naturally discover their own unique quality of being—or essence. This essence—this unique and crucial self—is at the root of persons' ability to develop and grow in positive, or pro-social, directions. This essence is "meant to be" from the beginning and, because it is unique, will continue to strive—even through adverse and troublesome times—to be fully present in the world. The concept of self-actualization maintains that there is a unique, positive, and mature ideal for each living person.

We are often amazed with the unique beauty of each tomato in our garden. If you slice one near where its stem was, there is a green star shape in the red tomato background. Each star shape is beautiful. Each is unique. Yet each can be recognized as the familiar star shape inside a tomato. While each whole tomato is unique, we can recognize when each has reached its version of a ripe, mature tomato. Nature provides many wonderful opportunities to reflect on how living things tend to strive for growth and maturation of "what was meant to be" from the beginning. In fact, it is almost impossible not to recognize the drive, determination, and resilience of many plants to survive and continue to exist in this world. Dandelions for instance—impossible to destroy—are constantly seeking out the conditions to fully develop into dandelions. Those irresistible fluffy round tops are constantly luring children (and some adults) to pick and blow—sending their multitudes of seedlings throughout the world. Just for the chance to get a wish! Those bright yellow flowers pop up everywhere, and actually duck when the lawnmower passes over them! How resilient and clever!

The concept of self-actualization maintains that each individual person's unique self—or essence—is likewise resilient, clever, and *impossible to destroy*. In multitudes of humans—despite the pressure to conform and live up to cultural, societal, parental, and other expectations—there is only one real you! And like the star in the tomato, your essence is beautiful and unique. You will continue to develop into a human being, for that is what you are. However, you are not merely the expression of a class of living things, nor are you simply the end result of your past experiences and genetic makeup. The concept of self-actualization maintains that within each of us is our own crucial and unique self—sometimes waiting or hiding or resting—but ever resilient and clever and wanting to grow in a positive, productive, and pro-social direction.

At one time we wondered, and it is often asked about in our classes, if each person is on a unique path to self-actualization, why wouldn't one, if given the optimum conditions and opportunity to follow one's unique path to reach self-actualization, actualize into pure evil? How can we be assured that self-actualization for some beings does not mean the allowance for the most perfectly developed evil beings? Are some human beings "rotten to the core"

Great?

or "bad seeds" from the beginning? Are some human beings meant to be evil? When a human being has committed acts of horrific violence and destruction, is this not an example of "the essence of evil?" Our answer to why we strongly believe this *is not* the case lies both in our spiritual faith and our beliefs in human nature and the power of therapeutic relationships. It is an answer we *continue to develop* in our experiences providing the optimum conditions and opportunities for our clients to self-actualize.

It may be most helpful to explain our deep belief in self-actualization by focusing on the apparent mechanisms that explain how this is. None of us live in isolation. Each person, even each living thing, interacts with its environment. Our environment seems to want us to reach our full potential and seems to prefer behaviors that in turn benefit the other members of our environment. For example, a human infant, if left without human touch and interaction, will fail to thrive. This is true even if the infant is fed enough for physical growth. To not provide care and nurturing for infants is unnatural to humans, so unnatural that, thankfully, infants are *in most cases* interacted with, nuzzled, held, cooed at, and protected by parents and adult caretakers who find this activity pleasing and mutually beneficial. Perhaps what is most pleasing is the very basic feeling of warmth involved in a caring human-to-human interaction. This caring behavior toward human infants is in turn beneficial to all members of the environment. Because we both work as counselors, we are aware that human interactions, when warm and caring, are almost always mutually beneficial. If either of us tended to be surly, harshly controlling, grouchy, poor listeners, and uncaring most of the time, people would understandably avoid us and we'd find ourselves alone, with growing unhappiness. Thankfully, these are not our usual characteristics and we find that others are often warm and caring toward us. While this explanation is overly simple, it is true for us. We find that the more we are warm toward others, the more others are warm toward us, the more we want to be warm toward others, and the more this exchange of warmth grows through our communities.

Mutually beneficial and supportive systems of interaction seem to be the way of nature. Picture a very green, open, meadow hill. As it is so green, it seems to be fertile and get plenty of rain and sun. One late fall day an acorn is dropped onto this hill, perhaps by a small animal. Through the winter and early spring, this acorn absorbs the moisture it needs and begins to germinate and sinks in as the ground begins to soften. Given its core conditions for growth (e.g., water, light, nutrients), that acorn will grow into a sapling.

If that sapling were truly alone on the hill, it might not develop to its full potential. Without the grass, its soil might erode away, taking needed nutrients with it. But as it grows and its roots grow mighty, it can then hold the soil for a greater diversity of plant life. Just as the bacteria and critters developed the soil that this sapling needs, its fallen leaves, the smaller plants its roots support, and eventually its decaying limbs and trunk will provide fodder for more bacteria and soil critters to thrive.

This sapling may one day develop into a mighty oak, almost as far across as it is tall. Such trees are visions of wonder and beauty. In maturity, they can take a rounded shape, in each season a different look and a different beauty. They keep the ground beneath them cool and moist through summer. They provide shelter from a storm. They provide homes for squirrels and launching pads for baby birds. Woodpeckers come to feed and keep the bugs that live on them to a healthy amount. They provide a place for a nap on a hot summer day. They provide a peaceful silhouette at sunset.

For us, such trees have become a symbol of self-actualization. They remind us that we all have the potential to reach our ideal. They remind us of the value of providing conditions for the full, strong growth of self in others. They remind us that allowing for expression, individuality, and the full self is good, and is beneficial and supportive to positive growth in others and the environment.

When Blocks Come into One's Path
toward Self-Actualization and Ideal Maturity

Imagine that lightning strikes that same oak when it is a young adult. That lightning may split off and deaden all the branches on one of its sides. Those branches may never recover. The tree may grow lopsided, instead of beautifully round in full strength. In a soaking rain and windstorm, it may fall over from this imbalance. It may die an early death, never reaching the ideal for its kind and not providing the full benefit it might to its environment.

For humans too, our paths to developing our potential may also be interrupted. Sometimes these blocks to full development come from great tragedy, from trauma or abuse. Sometimes these blocks come from a lack of the conditions and nutrients we need to grow and thrive (e.g., food, water, love, safety). Oftentimes these blocks come from a combination of trauma or abuse, a lack of conditions for growth, and our individual, human interpretation of these situations. A person's path to maturity can be interrupted, and thus that person's path to self-actualization may become much less than it ideally might have been. But there are differences between human beings and the oak tree. These differences give us a much greater opportunity to recover, to readjust our path to self-actualization, and to restart our striving for our ideal maturity. We illustrate key differences in the material that follows.

The Capacity for Awareness,
to Reason, Question, and Choose

A major difference between the oak in our example and humans is that we have a capacity for awareness, and abilities to reason, question, and choose. While it is true that our genetic makeup, our upbringing, and our life experiences certainly shape us, and in turn seem to force each one of us to choose a certain path, it is also true that we have the ongoing ability to choose alternate paths. These capacities and abilities allow us to recover and restart our path to our ideal maturity. We are able to feel fear and choose to avoid dangerous situations. We are able to feel warmth from others and seek greater closeness with them. We are able to feel dissatisfaction or angst, then to question and change our life's direction.

We are able to discern how some of our actions may leave us feeling, and make decisions based on this understanding. We are able to feel with others, to have empathy. From there, we are able to discern how our actions may affect others, to choose the effect we'd like to have and act accordingly.

We wish to note that these capacities and abilities may not be what we often think of as intelligence. While insight can be important, we do not mean to imply that it is all-important. Through recent research (Demanchick, Cochran, & Cochran, 2003), we have been reminded that even persons who have been consistently measured as having low intelligence (low enough to preclude insight, at least in the way that persons of normal or above intelligence tend to think of it) can benefit and change through therapeutic relationships and, through these relationships, restart their path to their full maturity.

Providing therapeutic relationships based on the skills of *The Heart of Counseling* is a way to prompt and to help persons recover and restart their paths to ideal maturity. Such therapeutic relationships focus persons on their awareness, reasoning, questioning, and choosing. Such therapeutic relationships provide a safe, sheltered environment for persons to restart and reshape their growth processes toward their full potential, toward the ideal person that their environment would like them to be.

Interpretation of Experience and Development of Self-Concept

We humans interpret our experiences and develop these interpretations into a self-concept that defines who we are and how we relate to our world. Then, quite often this self-concept becomes self-perpetuating as persons interpret new experiences in ways to fit their initial interpretations. Consider the following example that illustrates, albeit without some of the true subtleties of life, how this can happen.

Imagine there are two babies of different parents. Call them Baby A and Baby B. Consider that Baby A is born to loving parents. They are reasonably together people, they love each other, and they really wanted to have a baby and to be parents. So, when their baby is hungry, uncomfortable, or cries for other reasons, they work to meet his needs. He finds that his whole world is his primary caregivers, that they/his world will provide for him, and that he is safe and comfortable. We don't mean to imply that babies think in words and sentences, like adults, but expectations of our selves and others begin to form even in infancy.

Then imagine that Baby B is born of parents under great strife. Their lives are very distressed. They have their own problems that draw their attention away from their child. They have great difficulty loving and seem unable to consistently care for her. So, their baby learns to expect that her world does not provide for her needs, and that her world is not a safe and comfortable place.

Baby A grows into childhood being consistently well cared for and loved. Again, while not in words and sentences like an adult thinks, Baby A expects this. Baby B, on the other hand, grows into childhood without consistent care and love, at least without love being consistently communicated. She learns not to expect love, but to expect dislike, strife, and danger.

Babies and young children are egocentric—they believe that their world revolves around them. Truly, if they are to thrive, it is best for their world to revolve around them. Babies and young children tend to believe that the events in their life happen for or because of them. So, Baby A, now Young Child A, begins to think (again, we are using adultlike words to represent child thoughts), "Hey, the world [his world is the whole world to him] is a pretty cool place. When I need something, it appears. These people that I really like, they really like me. I must be a pretty cool kid!" But Baby B, now Young Child B, thinks, "The world [her world] is a dangerous place. I never know what's going to happen. People [her primary caregivers] don't seem to love or like me, and even seem bothered and upset by me. This must be my fault. I'm a terrible kid!"

Now Young Child A and Young Child B have begun to act on their beliefs. Young Child A greets his parents warmly and reaches to hug them. They hug him back and all are happy. He reasons, "See, it's true, I'm a Very Cool Kid!" Young Child B does not reach to hug others. Why would she? According to her expectations, it might not be reciprocated or might be met with strong negative emotions, having been just one more demand for over-stressed primary caregivers. Her belief system remains intact and is reinforced with her many lonely moments.

Now Children A and B go to school. Child A expects his peers and teachers to like him—after all, that's the way his world works. In fact, most of his peers and teachers do like Child A. Interactions with Child A are enjoyable and mutually beneficial. Most of those he meets are happy to return his warmth, to see his smiling face and his confident, pro-social behaviors. When there are a few who seem to not like him, he ignores their reactions as anomalies. Their reactions make no sense with what he *knows* is true of *his* world.

Child B, on the other hand, does not expect others to like her. She tends to scowl more than smile. She does not readily cooperate or follow instructions. Why should she? They don't apply to her. She is one thing and that world of others is another. So, with that attitude, her expectations are fulfilled. She is disliked and is isolated among her peers. When some teachers and peers do try to befriend her, she must ignore this or react with suspicion, as it makes no sense in the world she knows and expects.

When the school counselor becomes concerned for little Child B and invites her into a therapeutic relationship, the first thing she may do is work very hard to drive the counselor away. Rejection is what Child B knows and expects, and it is familiar to her. She has come to believe she can't be understood and won't be loved and liked. But in time, a consistent therapeutic relationship with her counselor begins to place useful seeds of confusion in her belief system. As the counselor comes to know her and feel with her, and to prize the core essence beneath her misbehaviors, she begins to think, "Nobody understands me, loves or likes me . . . except, well, maybe my counselor.

I've shown her everything. All my worst scowls and growls, and yet, I think she cares for me." This new thought chain is way too long for an automatic thought (a repeating self-statement underneath fully conscious thought). Now Child B has a useful confusion that prompts her to reconsider what she believed of herself and her world, to reconsider the attributions, and interpretations she has made of it. Her school counselor provided her with a safe place to stop, feel, and think, and to become aware of who she really is and what her relationship with the world can be. She provided her the opportunity to restart and renew her path to ideal maturity and her full potential.

Awareness of Existence,
Choice, and Questions of Self-Worth

We humans have the capacity to be aware of our existence and the opportunity to define ourselves. Toddlers go through a phase of asserting their existence, autonomy, and choice through saying "No!" to their parents, often asserting "No!" over activities that they are not actually opposed to. Young children continue defining their selves by defining their likes and dislikes, picking favorite colors, animals, and foods. Teenagers define their selves by which groups of friends they identify with. Young adults continue to assert their definitions of their selves through choices in clothes, hair care, and activities of interests. Throughout our adult lives, we continue to define ourselves through our lives and occupational choices.

We see these and the multitude of human choices as a result of our awareness of our existence and as ongoing efforts to answer the core questions of "Who am I?" "Who do I want to be?" and "How do I relate to my world?" We as humans also tend to continually value or devalue ourselves, by asking questions like, "Am I okay, likeable, lovable, worthy?" We attempt to answer these questions by considering how we relate to our world, asking questions like, "Do others like, love, or respect me?" "How do I compare to others? Am I better, stronger, faster, smarter?"

These questions are sometimes asked in full awareness, but are more often buried deep in our minds, under layers of other thoughts. We hear our answers to these questions in our minds over and over, and they affect how we feel and act, especially when we do not fully realize them. A metaphor for how this can be is the phenomenon of people working out in gyms. During most workouts, our minds are occupied with counting repetitions, miles, and minutes, or with other surface-level thoughts. Many persons who work out in gyms like to have music playing that is energizing, motivating, or calming. If the music was suddenly stopped and patrons were asked what song was playing, most would not know. Yet, most would agree that the gym's music has an effect on how they feel and what they do. Our self-talk answers to the questions of our identity and worth are much the same in that we often don't consciously hear them, but they are there, playing over and over, affecting our surface thoughts, our feelings, and our actions.

As you invite clients into therapeutic relationships, based in the skills of *The Heart of Counseling,* you provide an atmosphere, a safe place, and set of interactions that bring thoughts of self and self-worth to the surface for conscious consideration. In their therapeutic relationships with you, your clients will tend to shift from evaluating their worth in comparison to others to valuing their worth based on who and how they want to be. This work helps your clients reevaluate their answers to core questions of existence and helps them take responsibility for how they choose to define their selves.

Self-Responsibility Is Anxiety-Provoking

With awareness of self, comes awareness of choice. With awareness of choice, comes awareness of responsibility. This responsibility can seem like an awesome power and persons may feel inspired. It can also seem like a heavy burden and persons can feel great discouragement and fear. In realizing that we own the ultimate responsibility for our selves, which are made up of our choices, we humans often feel great anxiety. We will often give away our responsibility for our choices to others in order to avoid the anxiety that often comes with responsibility for our selves and our actions. In the United States, majorities of citizens do not even vote, much less take an active role in our society's governance. Yet, those same persons who give over this responsibility will also often complain about our society's governance. The game there seems to be, "If I take no action, resulting effects are not my responsibility. Therefore, I am off the hook and can avoid the anxiety of responsibility and choice." Unfortunately, not choosing is a choice. If citizens do not choose, others will choose for them, and therefore we are each still responsible for the outcomes in our society. *So true*

In spite of the discomfort of anxiety, when fully faced, experienced, and owned, the anxiety of responsibility and choice is very useful in helping us humans to reach ideal maturity. When fully faced, experienced, and owned, it sharpens our faculties and opens our hearts to make our very best, most satisfying decisions. The therapeutic relationships that you provide will help guide your clients to a safe place where they can face, experience, and own this anxiety. This then allows them to begin to make fully responsible and optimally satisfying choices. At the same time, through your acceptance and empathy in these relationships, your clients can come to be motivated by, as well as at peace with, the reality of responsibility and choice. Certainly, a desired outcome of therapeutic relationships is not to make this anxiety go away, but to harness it. A desired outcome is not to make decisions for clients, but to enhance and empower each client's own process of anxiety leading to choice and responsibility. *me.*

Awareness of Aloneness

Along with awareness of our responsibility and its accompanying anxiety, at some level, we humans also become aware that we are alone in our choices. We are alone in the sense that while many can help, no one can live our *counseling* lives for us or take the responsibility of our choices from us. This experience of aloneness drives us to relationships with others. This experience of aloneness is one of the elements that give great power to the deep connections of therapeutic relationships.

Emotions Are Useful and Necessary for Growth

While we may fear, avoid, or at least wish for painful emotions to go away, each is a useful, purposeful opportunity. Although it is an oversimplification, please consider our analogy between the usefulness of emotional pain and physical pain. If a momentarily unattended young child touches a hot burner, that child learns to be more careful around the stove. If that child didn't feel the pain of touching a hot burner, that child might not know to pull its hand back before serious injury. If a baseball pitcher develops a pain while throwing, that pain can be an early warning to make postural changes when pitching, in order to avoid a lasting injury.

We see emotional pain similarly. When it occurs, it has reason. When a person feels emotional pain after the breakup of a romance, there is opportunity through that pain, if fully experienced and contemplated, to decide and be motivated to make changes in one's self or actions for future relationships. When a person fully experiences the pain of a loved one dying, numerous lessons of how to live, how to care for one's self and others, can grow from that pain.

We have known clients who have come to counseling feeling emotional pain, anxiety, or depression that they did not understand and that could not be easily explained. The natural inclination of persons in such situations is to try to make the pain go away. But to do so may be to lose the life lessons that the pain offers and to risk greater pain later. Jeff remembers one client in particular who was brought to face unexpected decisions and responsibility through unexplained pain.

> He was living what he'd thought was a perfect life. He did well in school. He had good job prospects. His family loved his fiancé. But after learning of a friend of a friend's sudden death in an unexpected accident, he couldn't shake his feeling of anxiety and developed ongoing difficulty finding motivation for tasks he'd enjoyed so much before.

> He asked me to help him make the pain go away. I offered him skills for managing or coping through painful emotions, but also offered him the opportunity to fully experience, explore, and learn whatever there may be to learn from his feelings. While it sounded odd to him, he did accept and engage in a therapeutic relationship with me. Through the process of our therapeutic relationship, he decided that he'd been trying to live the choices that he thought others wanted him to make and that his pain might be telling him that it was his life to live and that *he* was responsible for his choices. (These are not words he actually said, but how I remember his discoveries.) He didn't end up deciding to break off his engagement, as he decided his love for his fiancée was true, or immediately change his career path. He decided, at his time of crossroads and life changes, to end his unexamined life and begin a life that was both full with his consideration for others and his consideration for himself.

You may note that this notion that all emotions have meaning and purpose also comes from or matches with a belief that all life has meaning and purpose.

For example, suffering may have many purposes, including teaching strength and appreciation of joy. If there is such a thing as evil, it may have purpose in defining good. As the adage from Philosophy 101 goes, "You can't make an omelet without breaking a few eggs." We hope our use of that expression doesn't sound trite or flip. Rather, we have a great respect for and place a great value on the suffering that is part of life. In accepting that suffering is *a part of life,* we also understand our responsibility to not *take this part from the client.* It is not ours to take away, avoid, or "talk the client out of." Because we want as helpers to care, to soothe, and to take away pain, this allowance for our clients to fully experience anxiety and suffering is very challenging, and is often where we counselors fail to trust in the power of the therapeutic relationship. Providing a therapeutic relationship will always involve deep caring, respect, and empathy for the anxiety and suffering of another human being.

Every Action Is a Choice of Destiny

As we write, we choose to be writers. In this moment we choose to work to communicate concepts that are very important to us and to our works to you in the hopes that they will be important to you and in your works.

Whenever we think of this concept—that every action is a choice of destiny—we are reminded of the many "choice moments" we have shared with children in child-centered play therapy (Guerney, 2001). Years ago, during a training session with other play therapists, we viewed on video one of our favorite examples of this concept (Nordling, 1998). A little boy in a play therapy session is exuberantly aiming and shooting around the room with a dart gun. He turns toward his counselor and he shoots a toy dart at her. Without missing a beat, or showing shock or disapproval, the counselor quickly and calmly acknowledges the little boy's wanting to shoot the dart at her, but then sets a limit (with empathy) to his behavior of shooting her. With just a moment's hesitation, the little boy seems to contemplate, but then with a little grin on his face, points the toy gun at her again. After reflecting his choice to do this, the counselor informs him calmly but matter-of-factly that if he shoots the dart at her again, his play session will end for the day. The little boy then deliberates for a very *long* moment. During this moment, the choice belongs completely to him, and he vacillates between pointing the toy gun at her, or at the ceiling light. He is making a choice for that moment, shoot counselor/end session, shoot at the ceiling light/continue session, express emotion by hurting/defying another, express emotion without hurting/defying another. Yet, he is also making a choice to define himself. Using adultlike thought sentences to depict his thought process, he seems to consider, "Will I be hurtful and defiant of one who holds me in positive regard and understands me, or will I be a person who would not do that?" "Do I choose to end my special time with this person . . . *I can choose.*" After his long moment of deliberation, it is clear, as he chooses to shoot up at the ceiling light instead of at the counselor, that *his seemingly small choice of that moment* is also a choice of his destiny, a choice defining who he is and who he will be.

While it is not always so simple or clear, our every moment and our every choice is a choice of our destiny. We define and redefine ourselves with the choices of each moment. The therapeutic relationships that you provide, based on the skills of *The Heart of Counseling,* focus your clients on their choices of destiny and the personal meanings that each choice brings.

The Internal World

The skills of *The Heart of Counseling* are focused on clients gaining mastery of their internal world. It is a pitfall for caring counselors to focus on tragic or unfortunate circumstances of clients' lives, over which clients have little control, versus the choices, feelings, and thoughts in response to those circumstances, over which clients have much greater control. The boy in the previous example may have had a very chaotic or abusive home life, and his counselor may have done what she could to bring about changes in his home life through consultation with the significant adults in his life. However, in his sessions, for him to make optimal progress, the focus should be on who he is, what he feels and thinks, and the choices he makes in the moments he is making them. Further, if his counselor focuses her thoughts, even after his sessions, on the troubles and deficits of his life, she will likely become discouraged. Instead, through providing a strong therapeutic relationship, she can tap into his resiliency and return him to his home environment having become stronger and more solid in his choices of who he is and who he wishes to be, regardless of how his environment changes with him. By dwelling more on her works to strengthen his resilience and choices, while also influencing his environment when possible, his counselor can both feel more encouraged and be more effective.

Locus of Control and Evaluation, and Being

It is desirable to shift our human locus of control and evaluation from others to ourselves. Because we have the capacity for empathy and because it is natural to desire relationships with others, if we live our lives based on our wants, we will satisfy ourselves *and* be considerate of others, rather than hedonistic. If we focus on valuing ourselves by our being, rather than valuing ourselves by how we compare to others or how we think others value us, we can make the clearest decisions of who and how we want to be and how we want to treat others.

On the other hand, if we strive to shape ourselves based on how we think we compare to others or based on the person we think others want us to be, we will develop a misshapen external self that does not fit our internal self. Then, attempting to maintain that mismatch and hide that assumed unacceptable internal self requires a constant tension, a great expense and waste of energy, often leading to premature stress-related illness, and limiting the potential possible in a congruent, self-actualized life.

The last underlying concept that we introduce is the belief that our human value is in being. We will never be able to count our value through our accomplishments or attributes. They are too tough to measure. There is always someone better. If we are unloved or disliked, even by those who we love, we are the same person. Our worth is always the same. This fits well with many spiritual beliefs (i.e., we are all part of the same creator or creation). It fits very well with ours.

However, such an internal locus of control and valuing based on being are very difficult to teach. Fortunately, they don't have to be taught. Rather, therapeutic relationships provide an environment and experience in which they are naturally learned or rediscovered. In such relationships there is nothing for clients to do but face their selves. If there are unsatisfactory behaviors, they will be realized and changed. If there are unrealistic expectations, they will be realized and adjusted. In therapeutic relationships, there is no one to defer evaluation to. In such relationships, there is no one to compare to, as the focus is empathically and acceptingly on one's self, alone.

ACTIVITIES AND RESOURCES
FOR FURTHER STUDY

- Remember to continue to consider and refer back to these concepts as you study and implement the skills of *The Heart of Counseling.*

- Discern our explanation and rephrase in your own words why, if each person's path to self-actualization is unique, some would not naturally develop to pure evil. Then, given this explanation, explain why some persons do develop into bad behavior at times in their lives. From what you know so far, explain in terms of self-actualization and other key concepts from this chapter how therapeutic relationships can help.

- In Skill Support Resource A (see back of book), we provide very abbreviated notes on each of our 11 concepts. We also suggest at least one initial implication for counselor action from each. Create your own examples and fuller explanations for each now and continue to add to them as you study and work. See what other implications occur to you. Revisit this activity throughout your development and exchange your ideas, examples, explanations, and implications with your peers.

- With a whole class or smaller group, take turns and allow each person to state what parts of these concepts make particular sense to you, what parts make no sense at all, what emotions you felt in reading them, and what related thoughts you had. Be sure to listen and accept each view expressed. After a view, thought, or feeling from one person is heard and accepted, another person's view, thought, or feeling may be added, whether or not it connects, differs, or matches the view, thought, or feeling previously expressed.

- Follow a similar procedure over a meal or other refreshments shared with one or more interested peers.

- Take yourself to a place where you can feel comfortable and calm, perhaps outdoors, under a tree, on a comfortable day. With a clock or watch in your field of vision, contemplate each concept for a few minutes. Begin by reading or remembering each concept, then let your mind flow from the concept to whatever memories, related thoughts, or feelings occur to you. Don't censure your thoughts or feelings; none can be wrong. When you realize your thoughts have drifted quite far, return your thoughts to the basic concept and let your thoughts begin to wander again. After spending a few minutes associating thoughts with a concept, move on to the next concept. Please note that we ask you to focus on both those concepts that appeal to you *and* those that *don't*. The point of this contemplation is not for you to convince yourself to agree, but if the disagreement between your views and ours (or at least some of the ways that we stated ours) has struck something in you (i.e., irritation, boredom, or strong judgments like, "That's stupid"), then there is even more reason to contemplate. This could be a very fertile moment for your learning. Perhaps in this contemplation, you'll come to an unexpected new thought that is particularly meaningful to you. Perhaps you will learn something new and useful of yourself and your views.

- Think of, draw, and/or build a symbolic depiction of each concept that will remind you of it and its meaning to you.

- Reread our tree metaphor, and then contemplate a beautiful, full tree in your life. Contemplate its perfection just as it is, and describe how it contributes to its environment and how its environment contributes to it.

- Revisit the Focus Activity of this chapter. Consider what you might change of your answers now that you have considered our views.

- Review the Primary Skill Objectives of this chapter, checking that you have mastered each to your satisfaction at this time. Reread, contemplate, question, discuss, and repeat further study activities until you have mastered each to your satisfaction at this time.

2

The Rich and Subtle Skills of Therapeutic Listening

We get into and out of difficulties with each other to a large extent by the way we listen and by what we hear, mishear, and fail to hear.

G. T. BARRETT-LENNARD

Talk low, talk slow, and don't say too much.

JOHN WAYNE
ADVICE ON ACTING

Man, if you gotta ask you'll never know.

LOUIS ARMSTRONG
REPLY WHEN ASKED WHAT JAZZ IS

PRIMARY SKILL OBJECTIVES

- Understand the complexities and difficulties of therapeutic listening for beginning counselors, for all counselors, and most especially for you individually.
- Be able to listen to another person speak on a topic that has at least a low level of personal, emotional content for at least a few minutes with nearly full concentration and focus on what the speaker is telling you.

- Be able to correctly summarize the core content of what the speaker has said within a couple of minutes, without missing core content and without adding any of your judgments or opinions, even if meant to be helpful.

- Tune out the thoughts in your head so that you can tune in more fully to the speaker.

- Identify your listening body language and several words that describe how it feels to *be* you physically and emotionally while listening.

- Be able to consistently identify when you and the listeners you observe are missing important parts of a speaker's communication and adding things that were not communicated, especially concerning the listener's thoughts and judgments.

- Be able to consistently perform or avoid nearly all the items on the Do's and Don'ts List for Therapeutic Listening while applying the skills in real-time practice sessions.

- Be able to explain the broadness of reflecting and create varied, realistic examples that you know of or can imagine.

- Be able to explain the connection and how taking the broad view of client communication can help you understand and implement a broad view of reflecting.

- Be able to explain what we mean by saying you are not, as counselor, a "paraphrasing machine."

- Understand how listening therapeutically is quite different from listening in everyday life, and experience some of the power inherent in listening this way.

- Explore some common difficulties in beginning to listen therapeutically and anticipate some of the difficulties you will face.

FOCUS ACTIVITY

In a group of three (better as classmates than friends, romantic partners, or family members, with whom you have an emotional connection and vested interest), have one person take the role of speaker and talk to you for three minutes on a topic that has at least a low level of personal, emotional meaning or content. Strive to commit your full attention to what the speaker is telling you. Try not to think of what you may summarize in response.

At the end of the speaker's three minutes, summarize what you happen to remember hearing. Strive to summarize only what the speaker said to you. Do not add your thoughts or judgments, even if you mean them to be helpful. Strive to summarize all of what you heard from the speaker, without leaving out any significant parts. Obviously, in this role, you are the listener.

The third partner, the observer, is to also listen to the speaker as well as to the listener's summary in order to help the listener know what errors he has made in summarizing the speaker's communication. While the speaker and listener are having their interchange, sit at least a few feet back to minimize any distraction you may present. Also, keep your eyes down or in the direction of the listener. This is to keep from encouraging the speaker from trying to communicate to the listener and to you at the same time—a natural inclination when two are listening. While hearing the listener's summary, prepare yourself to give feedback to the listener, including what he summarized well, missed, seems to have heard wrong, and added, especially any of his own thoughts or judgments, no matter how well-meaning. We do not encourage the speaker to give feedback immediately following this first practice, as it is so very tempting to add to or change what was said. As observer, you may keep time for the speaker and listener. The observer's role can be filled by either one person or a small group of persons.

Rotate through speaker, listener, and observer roles until everyone has had a turn. After each interchange of the speaker talking for three minutes, the listener summarizing for two, and the observer giving feedback to the listener, you may want to take a couple of minutes to discuss what seemed easy and difficult in the listeners' roles. We encourage you not to divert into the content of what the speaker said at this time. Do not use this discussion time to offer the advice or related thoughts that you may have wished to say to the speaker. Dismissing such thoughts when listening is one of the difficulties of therapeutic listening. Expecting to say them later will only divert you from listening, when it is time to listen.

After each has had a turn at all three roles, discuss the things you have learned that may be informative about what counseling is like, both for counselors and clients.

TUNING IN AND LISTENING

We assume the focus activity was difficult. As Schuster (1979) wrote of the difficulty of listening well, "Although this requirement appears to be extraordinarily easy to accomplish, in reality it continually slips through our [trained psychotherapists'] fingers" (p. 71). While many of us think we listen well in our lives outside of counseling, good therapeutic listening is extremely rare. To illustrate this, we sometimes ask how long family, friends, or acquaintances usually listen, before asking a question (which usually suggests a solution), directly suggesting a solution, or shifting the conversation with a "me, too" statement (such statements usually start something like, "I had a situation just about like that . . ." or "When I was your age, I . . ."). When we ask this of students and others, the answer we usually get is lightly stunned looks suggesting such listening time is quite rare in their daily lives. We often guess 10 minutes, and are told the time is not near that long, but more like zero to a few minutes.

Listening is a huge percentage of what counselors do. Therapeutic listening, as described throughout this chapter is profoundly different from listening in normal social interactions, and it is not easy. It is the first skill set of a developing counselor.

In this chapter, we focus on the skills of listening and reflecting what you understand of each client's communication, situation, and person. These skills demonstrate to each client that you want to know her, that you are striving to understand with increasing depth, and that you are beginning to understand and know the depth and complexities of the person of each client. In coming chapters, we focus more on expressing empathy, which communicates your deep understanding, feeling with, and personal connection to each client. Therapeutic listening is preliminary to the higher goal of an empathic and deeply personal connection with each client.

THE MANY LEVELS
AND NUANCES OF REFLECTING

The Broad Skill of Reflection:
From Paraphrasing to Themes to Confrontation,
Challenge, Summary, and Beyond

As you continue your study of this chapter, you should come to see that reflecting sounds simple but is complex and is broadly encompassing. Briefly defined, _reflecting is counselors' attempts to communicate to clients that they are striving to understand, and perhaps do understand, them._ The simplest form is to paraphrase back what you heard your client say. Another simple form is when, without even trying to, your facial expression matches the tone or emotions of your client's communication.

Paraphrasing Note that while paraphrasing and short summaries are the simplest and by far most common forms of reflection, they are only a place to begin. They are the simplest way to communicate your striving to understand, when you don't understand much yet. Since they are tools to communicate your striving more than your understanding, we encourage you not to worry over perfect wording, but to strive to understand the person of each client and his communication and to say what comes to mind as you strive to understand. If you think too much of how to paraphrase your clients' communication to you, that thought process will take you away from a focus on a personal, emotional connection that is the greater key to the heart of counseling. Once you know a client more, move to higher and more complex forms of reflection, rather than focusing on longer summaries. Descriptions of more complex forms of reflection follow.

Themes One higher-level form of reflection is to reflect patterns or themes in your clients' communications. For example, a client has been telling her counselor about her romantic relationship. Clearly it has been on her mind, but she has assured her counselor that she is happy with it. So, the counselor has made simple reflections like, "So, it sounds like you are really quite pleased with your relationship." As the client continues to discuss the relationship frequently, her counselor may reflect the pattern, "Seems like your relationship has been on your mind frequently in our time together." (The counselor's tone is interested and neutral; she is not indicating that it should not be on her client's mind or that there is any implication given to the fact that it is.) If the client then adds, "Yeah, I think I think about it all the time. I guess I really have some worries . . . ," her counselor might then respond, "Oh, so you see your relationship in a positive light, but now you've begun to have some concerns." (The counselor's tone would again be neutral, rather than perhaps ✗ excited that now we have a problem to talk about.)

Reflecting Discrepancies or Confronting and Challenging Reflections can cover complexities of communication and contradictions. One example of this can be, "So while you are happy with your relationship, it seems that was a little hard for you to say, and I thought you looked a little unsure when you said it." Another possible example is, "I get the idea that there is *so much* you like about him, and you are very glad to be with him, *and* you also seem to experience some doubts and worries." Another example could be, "A part of you wants very much to be with him, *and* a part of you wants something different."

Some counselors think of this level of reflection as confronting or challenging. Our thought is that such reflections are only confrontations if you have stopped striving to understand your client and mean to make a point (whether you are aware of your intention or not). For example, a counselor who is confronting may be pointing out that there is a contradiction within his client and there should not be. That counselor may be implying that the contradiction he has discovered is particularly important for his client to consider—that he knows best what his client should communicate next. Or a counselor may imply, at least in tone, support for one contradictory inclination of a client over another. Such errors mean the counselor has stopped listening and started telling and tend to undermine harnessing each client's drive to self-actualize and the development of self-responsibility.

We guide you to keep your focus on listening and respecting each person's self-actualization process versus making a point about a contradiction that you notice. Your reflections will include a contradiction that you see, but your purpose should remain saying what you see/understand of the person. This may include reflecting it when you see that your client's statements contradict each other, that her words contradict her actions, or that her words contradict her body language. Discrepancies within persons are part of being human. So reflecting discrepancies when you see them need not be confrontational. Regarding the notion of challenge, we find that reflecting contradictions tends to be naturally challenging, as we humans tend to want to resolve internal conflicts in order to feel better and to be understood.

I always forget... so much to remember

Regarding Summarizing or Saying What You See of the Person of Your Client We urge you to let your summaries occur naturally as you come to understand the person of your client versus initially only understanding little bits of communication, say the larger understanding you have of the person of your client and of that person in the context of her situation. Deep understanding of the person of your client in her situation is more important that summarizing content. Your focus on understanding the person of your client allows your summaries to be of the person rather than of the information she has given you; such summaries are naturally flowing rather than contrived. If you work to force a summary into your reflections, it will shift you to thinking about the person's communication and away from the person and the emotional connection that is more central to the heart of counseling.

To carefully let your client know how you see her is a broad and advanced level of reflection. In an exasperated moment your client might say to you, "It's just that I'm so selfish!" Then, if it is true to your perception of her, and you have begun a strong therapeutic relationship, the following interchange might occur:

Counselor: You see yourself as selfish and you *do not* like that about yourself.

Client: Yes, it's *so* true. Isn't it terrible?

Counselor: I hear that it's terrible to you. I hadn't seen you as selfish. I had seen you as being very careful not to hurt others.

The Broad View of Communication from Clients

While realizing the broad nature of reflections, it is also helpful to see the broadness of client communication. Everything your clients do or say to you is a part of their expressing who they are to you. This includes your clients' words, tone, facial expression and body language, pauses in speech, and even what they do not say. From the previous example, if the client has discussed her relationship in each session, then one week says nothing about it, you might reflect, "I notice the subject of your relationship does not seem to be on your mind today." (Neutral tone—the counselor is interested, but not suggesting the client should talk about her relationship.)

A young woman was telling Jeff about difficult family situations regarding her mother's illness.

It became very clear that she was talking about or all around her fear of her mother's death. I reflected something like, "I'm not sure, but I get the idea that you mean that this is so important to you, because you fear your mother may [slight hesitation] die." She paused, seeming stunned, and acknowledged that this was very true before continuing with added awareness. When I made that reflection, I was very tentative because I realized as I said it that this was not something that she had quite said yet. I hesitated before saying the word *death* because I feared it would be hard for her to hear, and because it was a hard thing for me to realize and say openly, as well.

This serves as a reminder that we are not listening machines. Therapeutic listening goes on, or in a sense exists, between two people, not between one person and one "paraphrasing machine."

In another example of the broad view of communication from clients, you can see that questions from clients are also statements about themselves. In the brief vignette at the end of the previous section, the client asked her counselor, "Isn't it terrible?" Rather than answering, "Yes it is terrible" or "No, it isn't terrible," the counselor took the question as a statement and reflected, "I hear that it's terrible to you."

In another example, a client told of a violent fight among her family with a very sweet smile. Her counselor reflected, "I gather this was *very* upsetting to you. I notice you smile while telling me about it." She responded, "Yes, that's what I do. I try to just keep smiling. If not, it would be too overwhelming." For another example, if a client consistently describes wanting a particular change, but consistently hesitates from actions toward change, you might reflect, "You've made it clear to me that you want to make this change, but something seems to be preventing you from the move you want."

THE DO'S AND DON'TS
OF LISTENING THERAPEUTICALLY

Overview of What You Are
Communicating with Your Actions

With the behaviors in the "do" list, you are communicating, "I understand what you are telling me." Even more basically, you are communicating, "I am listening," "I am striving to understand."

Do's

__ use your personal version of listening body language

__ reflect your perception of what your client communicates

__ make your reflections declarative statements, when sure that you understand

__ make your reflections tentative declarations, when you are unsure if you understand

__ keep your reflections short, whenever possible

__ focus your reflections on your client's main point, or the things communicated that seemed most important, most emotionally laden to your client

__ be prepared for and accept corrections

__ interrupt your client carefully to make reflections. Considerations for the counselor here are: (1) Will an interruption to reflect help the client clarify communication, thoughts, feelings? and (2) How much

helpful - feedback from practicum

communication can you reasonably reflect in a short paraphrase, before becoming overwhelmed?

__ allow your client to own most silences

Don'ts

__ allow interruptions for reflections to set up a hierarchy where your communications would seem more important than your client's

__ ask questions, except in rare circumstances, or state reflections in a questioning tone

IMPORTANT EXPLANATIONS
OF THE BEHAVIORS OF THIS
DO'S AND DON'TS LIST

While we use the Do's and Don'ts List to help simplify counselor behaviors, such lists oversimplify the complex and subtle processes of counseling. For example, the list given here and this chapter focus only on therapeutic listening. While this alone can be powerfully helpful, the much more powerful skill that we are building up to in our explanations is combining therapeutic listening with expressions from a deep empathic connection.

Your Listening Body Language

We emphasize *your* in this item because body language skills are unique to each counselor. Listening therapeutically is a way of being, rather than a set of fixed behaviors. This is also true for the body language of the listener. When you are listening, tuned in, and experiencing empathy, it will show in your body language. A story from Jeff's early work illustrates how your listening body language can be quite individual, yet still effective if you are focused in your groove of listening:

> My listening body language may be peculiar. I tend to shift and move within my chair a lot. This caught my attention when watching a session tape in fast-forward motion. While my client was doing most of the talking, I was much less still than her. When I stopped to watch the section at regular speed, I realized that my mind and heart were absorbed in her communication. As I saw that her communication continually deepened, I could see that she perceived that I was listening, fully focused on her communication. The fact that I did not sit still was not a distraction to her.

Our recommendation for developing *your* listening body language is to be aware of yourself in conversations. Notice when you are fully focused on the speaker's communication and when you are not (i.e., perhaps bored or preoccupied with thoughts of your own). Watch your body language at these different times and how it shifts in subtle or blatant ways. Then videotape yourself in both role-play and real sessions. Again, watch for the changes in your body language when you are fully tuned in and when you are not. Use this knowledge of your listening and nonlistening body language to prompt yourself to listen well, and catch yourself when you are off track.

Consider these hints toward generic listening body language. Listening body language usually includes looking at the speaker, and frequently means looking into his eyes, but without staring. It is usually not a closed posture (e.g., arms crossed). In review, you should be able to see shifts in your posture that fit with what your client is expressing. For example, when your client is telling you something over which he is excited and wants you to fully understand, you will likely sit up or lean forward. When your client tells you, or indicates, that what he is saying is difficult and will take time, you may sit back, showing that you are settled in and ready to listen patiently.

Reflect Your *Perception* of What
Your Client Communicates

No counselor or person can truly know what a client or speaker has communicated. Yet, it is the counselor's job to understand what is communicated. In striving, we may come very close.

A key reason for never truly knowing what another person is communicating is a basic flaw of using language to communicate. Language is always culturally laden. Words and phrases have meanings defined in dictionaries and implied meanings defined by culture, tone, context, and more. On the island of Guam, the Chamorro phrase for thank you is, "Si yu'os ma'ase." This translates something like, "Thanks be to God." This illustrates a cultural difference in language use. In America, the predominant culture seems more highly focused on the individual. We say, "thank *you*." It may be that the predominant culture of Chamorros is more focused on the collective, so the phrase used for thank you is quite different. For another example, in English we say, "You're welcome." In Spanish, the phrase commonly used in the same place is "De nada," which translates something like, "It's nothing." Again, there seems to be a cultural meaning implied. "You're welcome," may imply "Yes, *I* have done something for you and *you* are welcome to it or deserving of it." The Spanish phrase commonly used in the same conversational place seems to say, "I would like you to think nothing of what I have done." Of course, the tone and context of all these phrases adds still *much more* to the implied meanings.

We very intentionally label this behavior "your perception." Your reflections will always be your perception, rather than *the* truth. Everything a speaker says must first pass through her own mental filter (i.e., how one sees the world and one's self), and then through the imperfect medium of language (i.e., often there are no perfect words to say exactly what one means—poets often work hours to years to select just the right words). Then what the speaker has said must pass through your (as listener's) mental filter as your mind automatically puts what you hear into your worldview. Then, this whole process must be repeated as you reflect your understanding of what has been communicated to the speaker. Whew! It's quite a cycle, and one of the reasons Nancy enjoys counseling work with young children who aren't so reliant on words to communicate in counseling! So, the cycle of each reflection *begins* with the speaker's meaning, but then must proceed through a process that looks like this:

> Speaker's meaning > speaker's mental filter > language > listener's mental filter > listener's perception of meaning > listener's mental filter > language > speaker's mental filter > speaker's perception of what the listener said

Therefore, we often use such phrasings for reflections as, "I hear you saying . . . " and "It sounds like what you are saying is. . . . " Beginning counselors tend to use these phrasings more. As counselors become more experienced and confident that they are close to understanding their clients' communications, they tend to use such phrases less.

Along with the development of your listening body language, you will also develop your idiosyncratic body language that expresses your level of understanding. For example, your furrowed brow might show your strain to understand. Your relaxed face and firm tone in speaking might convey your certainty. Your tentative tone in speaking might convey your lack of confidence in your understanding so far. However, don't let your tentative tone have you drift into a questioning tone. We suggest you preface your reflection by stating that you are unsure rather than shaping your clients' communication with inadvertent questions. For example, you might reduce your tentative tone by prefacing your reflection like the following, "I'm not sure I understand yet, but I think you're telling me that what is so hurtful about it to you is the disrespect you heard in his tone."

Reflections as Declarative Statements

When the speaker's meaning is quite clear to you, your reflection should be a clearly declarative statement. For example, a client was telling how he had given directions to a coworker. He explained that he had spoken slowly, given the instructions a couple of different ways, and that it was really a simple task. He concluded emphatically, "So, I have told him and told him and told him, but still he doesn't get it!" A declarative reflection, over which the counselor could likely be sure, is, "You see that you definitely explained it well." A tentative declarative addition, over which the counselor might not be quite sure, would be, "And it seems to make no sense to you that he still does not understand." If this speaker then continued in the same vein, with emotions rising, and a look of irritation and disgust on his face, you might make an even more tentative reflection by saying, "I get the idea that you are *irritated* with his not understanding, maybe even disgusted."

The previous example begins to get into the realm of emotions. This is a most important topic and will be the focus of the next chapters. For now, the point is, when your client's meaning is obvious, you may drop phrases like, "I hear you saying. . . ." An example of this is, "What hurt you was his tone of disrespect." When your client's communication is not obvious to you, make your reflection a more tentative declaration with phrases like "You seem to. . . ." For example, "You seem to have been most bothered by his tone of disrespect." When your client's communication is not clear to you, but seems quite important or emotionally laden to your client, make your reflection an even more tentative declaration, with a phrase like, "I get the idea that. . . ." An example of this is, "I'm not sure, but I get the idea that you heard his tone as disrespect and that is what hurt."

The Question of Questions or Questioning Tone

Skipping ahead to the second "don't" on the list (Don't ask questions, except in rare circumstances, or state reflections in a questioning tone), it is important that you not let your tentative reflections end with a questioning tone. In fact, we recommend that you use questions only in rare circumstances (e.g., helping a client who has trouble getting communication started; genuine clarifications when you truly have no idea of your client's meaning; or perhaps in managing situations of potential imminent danger, which are addressed later in this book). Questions or a questioning tone are usually focused on the counselor's rather than the client's need—for example, to satisfy a need to have perfect reflections, to not be seen as misunderstanding, or to apologize in advance for a possible failed understanding. In fact, *as long as the counselor's intent is to understand the client, there is no penalty for being wrong with a reflection.*

However, there can be numerous time and relationship penalties (at the very least time inefficiency) to questions or a questioning tone. Many clients will try to answer your question, whether it is what they need to communicate or not. A questioning tone usually implies a yes or no question (i.e., "Is this what you mean to say, _____?"), which tends to stop the flow of clients' communication. Further, questions often imply that you are investigating a perceived problem in order to prescribe a solution. The alternative of directing a client to "tell me more" carries with it the same problems as asking questions. It suggests to the client that *now you are onto an important topic, perhaps one that will be useful to me in prescribing a solution.* In reality, such solution-based prescriptions are rarely possible or helpful in increasing the emotional well-being and self-efficacy of clients. By many accounts we are excellent counselors, yet we cannot expect to successfully prescribe solutions for the ongoing struggles of our fellow human beings. If we could do this, we would. Then, we would be very rich, which so far we are not. Additionally, such problem-solving questions communicate to your clients that you, as the counselor, do not trust their judgment or care enough to have the patience to allow them to discern their own solutions, or even to define their own problems. We will continue to show you a nonquestioning approach to counseling that avoids the problems of questions and maximizes time efficiency.

interesting outcomes?

A vignette that illustrates some of the problems with using questions follows. We pick up the dialog midsession:

Client: My boyfriend is so mean to me that I don't know what to do.

Counselor: How long have you been together? [Counselor may have only wanted context to better understand what speaker was saying or may have been attempting to judge whether her client was in a long or short time relationship and therefore whether it would be easy or hard, or even right or wrong, to end it.]

Client: We, uh, we've been together for two months and we, uh, I . . . I, uh, I love him so much that we started you know . . . being intimate last weekend.

Counselor: You've been together two months and you decided last weekend that you loved him so much that you would be sexually intimate with him? [Questioning tone is denoted with question mark at end of statement.]

Client: [Although it may or may not be true, at this time the client believes counselor is judging her critically. She looks down and pauses before she speaks.] Well, he's really a great guy . . .

Counselor: [Interrupts to ask] But you said he was mean to you? [This can be the worst kind of question. It seems like an argument from a lawyer on TV to a witness who has changed her story on the stand. The "counselor" who asks this believes she already knows the answer. Thus the question is rhetorical, pointing out that the witness is changing her story and now may be lying.]

Client: Well, he really is; he has a great future, he's captain of the football team, all the girls love him, and he loves me.

At this point the client seems to be shifted to the opposite of what she came to say. She believes she must defend herself and her boyfriend from her counselor, whom she sees as disapproving. Their therapeutic relationship is in serious jeopardy. This client will not let her counselor know who she is while she fears critical judgment. As she cannot let her counselor know her real self, she cannot do much work on her real self.

However, let's assume a miracle happens and this client does get back to what she seems to have come to say:

Client: What I mean by his being mean is that, well, he's scary sometimes.

Counselor: Tell me more. [This is a directive that functions like a question.]

Client: It's just that before we did it . . . had sex I mean, he could be sooo rough, and even when we did he . . .

Counselor: Are you saying he is abusive?

Client: Oh, no. [Confused, afraid, perhaps defensive again.] I . . . I told you he loved me.

Counselor: But this is important. Is he hurting you?

Client: I . . . I don't know . . . no, he would never do that. He's from such a good family and his parents love me too.

Now, while this counselor surely meant to have her client's best interest at heart and obviously cares about how apparently weaker partners are treated in relationships, her questions and way of being with this client have produced a defensiveness in the client that may preclude, or at least greatly lengthen the time, it will take for her counselor to help her. Also, if this client is sometimes easily manipulated, perhaps due to self-talk that tells her she is weak or undeserving or must/should depend on others, and can't make good decisions for herself, this counselor's questions have reconfirmed such beliefs. If this client is in an abusive relationship, even if the counselor were able to later talk the client into leaving the relationship after judging that it was abusive, the counselor would effectively be asserting herself in the place of the controlling boyfriend. One cannot control other humans in order to have them come to control themselves.

If, on the other hand, this counselor had focused her energies on therapeutic listening, especially with expressions of accurate empathy, this client may have soon come to develop and trust her judgments of what she deserves, her understanding of how she is being treated, and how she wants to be treated. Consider the following small piece of how that work might go:

Client: I just don't know . . . he's a great guy, everybody says so . . . but he scares me, sometimes.

Counselor: [Tenderly, similar to the tone with which the client spoke.] You seem to feel confused. You've told me you love him, you know that others see him as a great guy, but you also feel scared by him, sometimes.

Client: Well, yeah . . . [tearing a little] he's the best guy I ever had [long pause]. But when we fool around or even do it, he hurts me. [Quickly adding] It's just little things, though.

Counselor: I got the idea you wanted to make it clear to me that it's just little things. It also seems you're saying they *are* things that bother you.

Client: [Very tentatively] Yeah, I guess . . . a little. You know, the other night when he was drinking with friends and I didn't get the joke, he, well he . . . he laughed at me [the pauses in this last statement were long and she was beginning to tear again].

Counselor: [Again tentatively, close to her client's tone] I gather from how hard that seemed for you to say, that that really hurt you.

Now, through therapeutic listening, this client is coming to express more of her relationship concerns to her counselor and importantly what she thinks and feels about her relationship. Through this therapeutic listening, she is coming to discover, perhaps admit to herself for the first time, what she feels. She is coming to value what she feels. She seems to be beginning to trust what she feels, as well as express and explore her doubts. Very importantly, she is coming to value herself, as her counselor values her enough to therapeutically listen.

*Good -
I were starting
to worry*

*I were starting
to worry*

     Also, if there is an imminent danger issue, it will most efficiently be uncov-
ered through this process. In Chapter 10 we will explain and illustrate how
counselors can assess for danger while maintaining therapeutic listening and
respect for clients and their thoughts and feelings.

Keep Your Reflections Short, Whenever Possible

Because you are less involved in your client's situation, you will often be able
to reflect your client's communication much more concisely than your client
was able to communicate to you. A rule of thumb for the percentage of talking that you do versus the talking your client does would be a quarter of the
communication by you and three quarters by your client. This rule of thumb
proportion is not something that you should concern yourself with much during sessions, as it would take your focus off your client and your client's communications. Rather, you should consider it when reviewing your work after
sessions. The shorter your reflections can be, while still capturing the truth
and not being hurried, the more they will enhance versus pull your client away
from expressing his experience to you.

The Special Issue of Verbatim Reflections

Verbatim reflections of client wording often make for the most concise, direct
reflections. However, verbatim reflections are sometimes not ideal, as they could
seem oddly echoing to clients. They are okay to begin with, especially when
they are said with empathy (focusing on the emotional meaning of the statement, rather than the words). Additionally, verbatim reflections are often just
the right thing to do, for example, when your client has said something that is
strongly emotional. When our clients state an absolute thought with strong
emotion, such as, "It's just impossible! He can't be worked with! [stated with
high frustration and exasperation]" it often helps them to hear their words back.
To this client statement, we would likely respond, "That's how bad it seems to
you. You see that it's *impossible* to work with him." In many situations like this,
our clients have responded something like, "Well it may not be impossible, but
I've had enough of trying [then decided to quit the relationship]." Others have
responded something like, "Well, it's not impossible. I have been thinking of a
way to change his mind [then proceeded to explore other possibilities]." The
point in making such a near verbatim reflection was not to question the client's
absolute thought (it should not be made with a questioning tone), but to let
the client hear her own words and tone to fully realize her experience. Whether
the situation is workable or not should remain your client's decision. Her best
decision will come about through the counseling process of providing therapeutic listening (especially with empathy and unconditional positive regard). It
is your role to focus on your client and her communications, without second-
guessing, and to trust the powerful process of therapeutic listening.

*helpful w/
Courtney*

     The following very useful application of verbatim reflection comes to mind
from Jeff's work:

> I was working with a client whose daughter was getting married about a
> six-hour drive away. His mother had health concerns that made it difficult

for her to travel and apparently seemed to sometimes manipulate and impose herself on her son (my client). He had made arrangements for his mother's travel and for her physical and personal care during the wedding. However, his mother refused those arrangements and insisted on traveling with her son to the wedding of her granddaughter. My client very much wanted not to have to take care of his mother during the wedding. He wanted to enjoy the event and be focused on his daughter. At one point he was so exasperated that he shouted, "God, I hate her!" So I reflected almost as strongly, "You just hate her!" My client then quickly changed his communication and explained to me that he could never hate his mother. I accepted this by saying, "Oh, I misunderstood. You would never really hate your mother." The next session, my client explained that he'd thought a lot about whether he did feel hate for his mother and that he certainly didn't want to hate her. These thoughts prompted him to set a firm limit with his mother; he told his mother that she could either accept the travel arrangements made for her or miss the wedding. My client explained that he had discovered he loved his mother, himself, and his daughter. While he wanted his mother to attend the wedding, he was most concerned that he be there to help his daughter if needed and to enjoy the time, especially since he was confident that his mother did not actually need to depend on him that weekend. He continued to find situations in which he had thought others absolutely needed him, but he came increasingly to find that they could also be okay without him when he needed time or energy for himself.

Focus Your Reflections on Your Client's Main Point, or the Things Communicated That Seemed Most Important, Most Emotionally Laden for Your Client

A common troubling issue for beginning counselors in approaching the 25/75 ratio is what or which communications to reflect. Your clients will often say a string of things in order to make one point. A counselor cannot reflect everything. You must choose from a client's many communications, and you need to choose very quickly, in a way that will come to be automatic and natural with practice. Your choices will influence the next communications from your client. Clients tend to say more about the parts of their communications that you have reflected. So, reflect the heart of your client's communication, meaning: Reflect the part of the client's communication that seemed most meaningful to the client, most emotionally laden.   *tough*

Don't let thinking about these choices take you off your focus on your clients and their experience. If your heart and mind are in the right place but you make a wrong decision, it will be inefficient, but there will be no harm to your therapeutic relationship. So you may wish to remember a quote that has often been helpful to our students and us: *As long as your intent is to understand your clients and their communication, there is no penalty for being wrong* (Eric J. Hatch, personal communication).

Be Prepared for and Accept Corrections

In the previous case example of the client traveling to her son's wedding, Jeff was corrected for stating a reflection that was almost verbatim. When corrected, it did not matter whether Jeff's statement had been actually correct or not. Whether you have actually misunderstood your client or your client thinks you have, the right, respectful, and efficient thing to do is to simply accept the correction. The correction becomes the client's communication. In the "going to the wedding" example, once Jeff's client told him that he could never hate his mother, that statement became the important piece of communication. To not accept this would have been disrespectful and would have meant that Jeff had stopped listening to his client.

The Issue of Interruptions

You will need to interrupt many clients in order to reflect. Many clients speak rapidly, moving from one topic to another without pause for you to reflect. Sometimes this may be because they are nervous and talk or ramble in a fast pace when they are nervous. Sometimes clients seem to do this because they are afraid that if they give their counselor time to think and respond, the response will be critical. Most clients don't pause for you to reflect because they don't have any way to know that that is what you intend to do.

Interrupting is sometimes difficult. We, and many of our students, were taught as children that interrupting is disrespectful. Yet, in counseling, if you don't reflect, it will be difficult for your clients to know that you understand or respect them. One of your primary functions in therapeutic listening is to help your clients more fully understand themselves and their communications, thoughts, and feelings. To do this, you must reflect.

Further, especially for beginning counselors, if you let your client communicate a great deal to you before reflecting, it may overwhelm your ability to listen, comprehend, and remember. If your intent is to briefly reflect your understanding of the heart of what your client has told you, your interruption will enhance, rather than stunt, your client's communication to you.

If you interrupt with an intention other than reflecting the heart of your client's communication, it may signify that you see your responses as more important than your client's communication. This would stunt your client's continued communication.

Assuming your client does not naturally leave you a space to reflect, then interrupt to reflect once a point is made that seems to have felt significant to your client. However, there are exceptions to almost every rule. One of Jeff's adult clients came and told him that she had remembered sexual abuse from her childhood. She made it clear, first with her body language, then with a direct statement, that she felt she needed to get the whole story out without stopping. He respected her decision for her use of the time and made very few reflections in her session that had been extended to an hour and a half.

A more common situation is the client who continues to talk rapidly, with few pauses. In such situations, ready yourself to gear up to this client's pace as much as possible, to make your reflections very short, and to make them while your client is still talking. For example, if your client is rapidly explaining multiple sides to a decision and seems to be describing himself as feeling torn or pulled in many directions, you may reflect as he continues to speak, reflecting single words or phrases with empathy and in his tone, like "torn," "so pulled," "you know you want that," "that is what you fear." Such sessions come to be as if you and your client are improvising music together. Your client provides the main melody, while you provide short phrases that repeat and underscore parts of that melody. When watching or listening to a recording of a session that you think is like this, you can check to see that your work is effective by noticing that your client does hear you, even though he does not stop talking, and that your underscoring parts of the melody help to further develop the song that you both are making, which is really his song, with you playing backup. One way or another, usually in easier ways than those described here, each client and you will develop a rhythm of communication and reflection that is natural and unique to the two of you together.

Allow Your Clients to Own Most Silences

There are quite a few good reasons why your clients may fall silent after you reflect during a session. They may be thinking of their next thing to say or how they want to say it. They may be girding their strength or nerve to tell you something very emotional or that is difficult for them. They may have been taken aback, surprised by the reflection they heard back from you. They may be taking a very valuable moment to sit with a feeling or thought, to experience it fully. They may simply, naturally communicate at a pace slower than natural to you. In these cases, it may have seemed to them that there was not a silent moment at all.

Counselors sometimes feel awkward with silent moments and either believe that their clients are much more uncomfortable with silent moments than they actually are and/or believe that they, as counselors, should be doing something to make the sessions smooth and to fill the silences. Remember that if clients fall silent after you have reflected, then it is not your turn to communicate again until they have. An easy, if overly simple, quotation to remember for this is, _the client owns the silence_.

It is also important to realize that the talk is only part of the communication in a counseling session. So, if in a silent moment, after you have reflected, your client looks at you with enlarged eyes and raised eyebrows, then shrugs his shoulders and turns his palms up to you, this is not silence. Your client has probably communicated something like, "You are not saying anything. I don't know what to do next, yet I think I should do something and I feel uncomfortable." To those highly communicative gestures, you could reflect something like, "We've fallen silent and something seems to feel uncomfortable to you about that."

Consider another example from Jeff's work:

I can remember a time when I was too quick with such an assumption. I was counseling a young man who was struggling with an important decision, over which he felt very anxious. He had put it off for a long time and was now under pressure to decide soon. We were developing into a rhythm in which he would tell me a part of his decision process or an aspect of his situation, I would reflect, then he would fall silent for at least 2–5 minutes. I misunderstood his facial expression and body language in an early one of these silences. My mistake was because of my awkwardness with silence and the thought that I should end it. It was also because I hadn't realized the emotional impact the decision had for him. So, I interrupted a silent moment to say, "We've fallen silent and I gather you are uncomfortable with that." The look on his face then turned to surprise and mild irritation. He explained, "No, I wasn't uncomfortable with *that*, I'm uncomfortable *with the decision*." That was a clear sign for me to learn to wait my turn. I came to understand the function of counseling for him around this decision as follows: Because he felt anxious around the decision, when he was away from counseling he avoided the anxiety by avoiding thinking it through. He came to counseling to discuss and contemplate the decision. In the silences, he was thinking and feeling about the decision fully. So, those silent moments were very much what he needed from counseling. We continued in our rhythm for several sessions while he made his decision.

HOW THERAPEUTIC LISTENING DIFFERS FROM LISTENING OUTSIDE OF COUNSELING AND FROM NONLISTENING

It is important at this juncture to revisit the fact that therapeutic listening is very rare outside of counseling. Loved ones, friends, and family usually listen to emotional communications only for fleeting moments, if at all, before attempting solutions, reassurance, or shifting the conversation with "me, too" statements. When loved ones, friends, and family discuss emotionally laden situations, there is often a fairly equal give and take, a sharing of experiences related to the topic. This can be very helpful. We often learn from the sharing of experiences and the guidance of significant others in our lives, but that is not counseling.

HOW AND WHY LISTENING THERAPEUTICALLY WORKS

It is partly the vast differences from conversing with loved ones, family, and friends that makes therapeutic listening therapeutic. The very fact that your clients will have met with you, a counselor and person, who responds differently to them from others who care about them helps them to begin to shift away from troubling habits, patterns, and ways of being.

Therapeutic listening prompts clients to focus on their situation, and their experience of it, rather than others' thoughts and feelings about their experience or others' stories of their similar, but different experiences. Especially when experiencing anxiety, we humans often avoid thinking through the anxiety-provoking situations. Anxiety can be quite uncomfortable, so we avoid it. This sometimes takes the form of thinking a lot about the situation, but unproductively thinking the same thoughts over and over. When addressing the issue of silences previously, Jeff gave the example of a young man who used therapeutic listening in counseling to keep himself focused on a decision that made him anxious—one that, outside of counseling, he worried over but did not think through or make progress toward.

Your therapeutic listening will let your clients know you are right there with them, even when they communicate and experience emotions that are very scary to them. As you listen therapeutically, your clients can and will take opportunities to face thoughts and feelings about themselves that they have avoided and kept hidden from themselves. As you listen therapeutically, your clients face these thoughts and feelings with true responsibility for themselves, yet with you. You are there with them, experiencing with them, yet not protecting them from their emotions and full experience, even if they fear their emotions, and even when the emotions hurt.

As in the example that Jeff gave under verbatim reflections, clients increase their self-awareness and insight through therapeutic listening. In that example, Jeff's client became aware of his strong irritation and of feeling trapped by his mother's behavior. His insight was that he was not satisfied with his relationship with his mother, that it was hurting other relationships, and it was possible to make changes. He made changes in his relationship with his mother especially, but also in relationships with others.

It is important to add here that while insight is often a by-product of therapeutic listening, it is not the goal and may be highly overrated. This is especially true if insight is misconstrued as analysis. An example of analysis as a mistaken goal of counseling came to us recently as a counselor strove to help his client understand that he was angry and that he lashed out at others because he was humiliated by his father. First, the client did not accept this thought from his counselor as true. He didn't agree that he was angry and lashing out at others. Further, he seemed to feel humiliated to have such an analysis suggested. Finally, even when accepting the possibility of the analysis, he still felt humiliated and still acted angrily.

We recently completed a research project that underscored the fact that insight (especially if referring to analysis) may be highly overrated. In this research, one of our former students provided person-centered play therapy to adults with developmental disabilities. Based on these clients' IQs, there was no possibility of cognitive insight, at least as many of us may think of it. The clients had minimal verbal abilities. Yet, through their counselor being tuned in to them, therapeutically listening to their communications through their actions and play, and through their emotional connection with him, they made behavioral and emotional progress that was important to them and to their caregivers (Demanchick, Cochran, & Cochran, 2003).

Therapeutic listening works because it is empowering. Through listening therapeutically, your behavior tells your clients that you value them and their experience, that you have faith in them and their movement toward self-actualization.

WHAT'S DIFFICULT ABOUT LISTENING—COMMON ERRORS AND PROBLEMS IN LISTENING FOR BEGINNING COUNSELORS

Common Interfering Thoughts

One of the basic difficulties is tuning in to the speaker. We must strive to quiet our minds during sessions. Often the voices in the minds of beginning counselors, who intend to listen, ask questions such as: What should I reflect? Is my client okay? Does my client like and respect me as counselor? Such thoughts may be inevitable in the beginning. In most cases when they occur, your job is to refocus your attention on your client and on listening to your client's experience. Following are some ideas that may help you with these commonly interfering thoughts.

What Should I Reflect? We gave our answer to what to reflect through our list of do's and don'ts (item: focus your reflections on your client's main point, or the things communicated that seemed most important, most emotionally laden to your client). When you are listening, this main point will automatically occur to you. So state what occurs to you and remember that *as long as your intent is to understand your clients and their communication, there is no penalty for being wrong* (Eric J. Hatch, personal communication).

Is My Client Okay? We will address the challenge of assessing for imminent danger and managing crises within your therapeutic relationships in Chapter 10. However, some brief guidance on this question follows, as it often interferes with listening. Remember that your clients are letting you know if they are okay or not okay in their communication to you. Listen and you will know. Then, when a client has given you reason to be concerned over imminent danger, let that client know your concern and that you are working to assess her safety while listening.

Does My Client Like and Respect Me as Counselor? This is not something counselors need to wonder about. As you listen to your clients, your clients will let you know their experience of counseling with you. For example, if your client seems eager to tell you more, then your client likely feels encouraged by your therapeutic listening. If after you reflect, your client gives you a disgruntled look, you may next reflect, "Something seemed to bother you about my last statement." In reviewing recordings of your sessions, you may learn that your clients seem to be making progress in moving toward the heart of what they have to say. For example, a client's communication may progress from speculations about the future to who he is now, how he is in relationships now, what he is not satisfied with, and how he wishes to be different in relationships. Another client's communication may progress from missing a deceased parent to feeling guilt when thinking she does not live up to what she knew her parent's expectations were and feeling proud when she does live up to them. Ultimately, it does not matter much whether your clients like and respect you. If counseling is particularly difficult for a client or a client does not yet see the value of counseling for himself, you may help him to hang in and work through the difficulty by explaining how his use of counseling can help him and therefore why you do what you do as his counselor. (We address this skill in Chapter 8.) While most of your clients will come to like and respect you, this is not a primary determinant of their progress. It's nice, but not necessary. The thing that's important is that your clients experience your genuine caring and concern (Cohen, 1994). Then, as you are consistently caring and concerned, your clients will come to know you that way.

Therapeutic Listening Is Not Normal and Can Feel Odd

Because therapeutic listening is very different from usual conversations, it can feel odd to beginning counselors. Some of our students who are beginning counselors tell us they feel awkward and out of control with the skills. This out-of-control feeling makes perfect sense. When listening therapeutically, each counselor is in self-control but is not in control of the conversation and content of the session. That control is given in trust to clients and in trust that the process of counseling will go where each client needs to go. This is a leap of faith on the part of the counselor. You have to do it and have it work to *know* it works. With effort, skill, and the right intentions, you will experience success and your role in counseling will come to feel familiar, comfortable, and natural.

[handwritten margin notes: "tough", "blind faith..."]

The Urge to Fix Immediately

Many counselors choose the counseling vocation out of a desire to help others. Many gravitate to it because they realize that others have frequently come to them with their problems and for advice. Thus, when faced with a client who is struggling or suffering, many beginning counselors are sorely tempted to give advice or otherwise take some action to lead their client to a quick solution. However, to give advice or attempt a quick solution is equivalent to admitting to yourself and your client that you have given up faith in the counseling process and in your client. Such directing is disempowering to your client and inhibiting to the therapeutic relationship that the two of you need to form. We encourage you to take the leap of faith of therapeutic listening, then to keep the faith and see it through.

Sometimes Clients Want a Quick Fix

Sometimes clients want what they expect will be a quick fix (e.g., advice or some other magical problem-solving technique). A first step in such a situation is to acknowledge (or reflect) your understanding of your client's desire. For example, "That's how difficult this is for you. You want a solution *now*!" or "That's what you expected from counseling, that you could give me a description of the difficulties in your relationship, then I could give you advice on how to handle it." Very often in hearing their experience reflected, clients withdraw their request for guidance. When such a request continues to be what your client wants, you may then need to explain or reexplain how she may best use counseling. (Example explanations of therapeutic relationships can be found in Chapter 7.)

The Desire to Ask Informational Questions
That Are of Interest to You

Beginning counselors often feel a temptation to ask questions of interest to them, for example: "Where is your mom now?" "How old were you when your father died?" or "What were the circumstances then?" Sometimes such questions help to convey your interest. More often they imply that you are gathering the information that you know to be pertinent in order to produce the best solution or course of action. We think of clients' experiences as compelling novels. While we may want to know the end of the novel at the beginning or may want to know certain details of characters' lives, we may not know until the author presents the information. In the case of counseling, the controlling author is both the client and the counseling process, but not the counselor.

Stories

In wanting their clients to make progress quickly through therapeutic listening, our students often worry that their clients are telling stories of others, rather than telling what their own experience is. We often tell students that everyone's stories are ultimately about themselves. As we tell you little stories of our interactions with students or clients, we are inadvertently telling you that we see ourselves as good teachers and counselors. We have an uncle whose every story is about how someone wronged him. We believe he is telling us that he sees the world as unfair to him. We have another relative who often tells disgustedly of others who are lazy. It seems she does not approve of laziness.

From our interests and expertise in counseling children, we know that highly significant parts of a child's counseling may be a battle in a sand tray between opposing forces or may be a dollhouse play with robbers and rescuers. In these situations, the counselor may never know just what the play says about the child. It usually does not matter. What matters is experiencing the story or play with your client and responding to the emotion of your client's experience. We will address this empathic connection through stories and other communications in later chapters.

Returning to Important Communications

Sometimes clients say things that seem very emotionally important to them only to move on to other communications before their counselor has the opportunity to reflect. Our students sometimes want to help their clients focus on the communication that seemed most important by asking/directing them to go back to the topic that seemed more important and to tell them more. While the direction may be useful, the way of doing it can be disempowering to clients. It is as if the client is communicating freely, following her impulses and intuition, coming to trust herself and this process, then her counselor breaks in to say through his question/directive, "I know better than you what is important for you to talk about. So talk about _____." A more empowering way to help a client go back to what seemed important to her would be a reflection like, "A moment ago when you spoke of your loss, it seemed important to you—you seemed to have strong feelings, then you changed the subject." From a reflection like this, the client may see that she does want to go back to the topic that was emotional, she may come to see that she routinely changes the subject from emotional topics, or she may correct her counselor in explaining something like, "Yeah, when I think about the loss it hurts, but the most important thing for me is to focus on what I'm going to do now." Whichever the client's choice and experience, it is the counselor's job to listen to that next communication.

MULTITASKING WITH THERAPEUTIC LISTENING

Every other task in counseling is in addition to therapeutic listening, never instead of therapeutic listening. In upcoming chapters you will add the skills of experiencing and expressing empathy; establishing and striving to maintain unconditional positive regard; working to ensure that your listening, empathy, and positive regard are genuine; helping clients get started; structuring and explaining counseling; assessing for progress; managing crises; and more. We hope this list does not sound daunting, but rather inspires you to practice therapeutic listening more often, both before and while adding more and more skills.

ACTIVITIES AND RESOURCES FOR FURTHER STUDY

- Now that you have read about and contemplated complexities of listening therapeutically, try the Focus Activity for this chapter again. Notice how the qualities of your listening have changed. Repeat the exercise until you are satisfied with your proficiency for that time and circumstance.

- Advancing from the previous activity, practice real-time therapeutic listening. Have a speaker (a classmate) talk to you about a topic that has at least some emotional content. Follow the Do's and Don'ts List and the guidance that you have read for therapeutic listening in this chapter. You may have one or more observers as well; they can give you feedback on your effectiveness and suggestions for improving. Begin with short practice sessions, 10–20 minutes, and expand to full-length practice sessions of 50 minutes.

- Journal or essay about difficulties in your development that you anticipate through practice, from readings, and from experiences with initial clients with therapeutic listening. Be specific and give examples of these difficulties, whether hypothetical or actual. Address what it is about you that seems to prompt these difficulties. Explain what you are doing and will do to work through these difficulties.

- Do the same for your anticipated and experienced strengths with therapeutic listening.

- Look back at the primary skill objectives and review or practice until you see that you have mastered each to a high degree and to your satisfaction. Perfection is not necessary and perhaps not possible, as you may always be working to perfect these core skills.

3

Striving for Empathy

Emotion is the chief source of all becoming-conscious. There can be
no transforming of darkness into light and of apathy into movement
without emotion.

CARL JUNG

When the eyes of the heart open, we can see the inner realities hidden behind
the outer forms of this world. When the ears of the heart are open, we can
hear what is hidden behind words; we can hear truth.

FADIMAN & FRAGER

I learn by feeling . . . what is there to know?

THEODORE ROETHKE

PRIMARY SKILL OBJECTIVES

- Develop a beginning understanding of what empathy is and what it is
  not. (A full understanding can come only through experiencing, receiving,
  and providing empathy, as well as ongoing contemplation.)

- Develop a beginning understanding of how and why empathy gives power to counseling and is a healing force. (A mature understanding of this power will come with experience, providing strong empathy.)

- Be able to explain the purposes and functions of empathy in counseling.

- Make significant progress in your ability to empathize. (Complete progress in this ability is a lifelong endeavor.)

- Identify what does, or may in the future, prevent your full empathy.

- Begin working to overcome the factors that prevent your full empathy.

- Research literature replete with observable, measurable evidence for empathy.

FOCUS ACTIVITY

You must experience empathy before you can express it. Attempts to use words to express empathy sometimes get in the way of beginning counselors' ability to experience empathy. So, odd as this activity for experiencing empathy may seem, we urge you to work through it.

You may either videotape or have a third partner observe this focus activity. Have a partner communicate to you on a topic that has at least some emotional content for him or her. You are to listen and strive for empathy. The speaker may use words to communicate to you. You may not speak in response. We urge you not to try to convey your empathy but also to make no attempt to hide it. Your task is to simply be with the person who is communicating to you and strive to feel what that person feels. We use the word "simply" to imply that this is your *only* task, not that it is an *easy* task. It may help you to be aware that empathy is not restricted to big, dramatic emotions. It is also used for the normal flow of emotions through all of us humans and can be seen in all of our communication and being.

HUGE

If you catch yourself thinking of a solution to your partner's problem while he or she communicates to you, you have strayed from empathy and listening. If you catch yourself thinking of similar situations in your life, you have strayed from empathy and listening. If you catch yourself thinking of the word for what your partner feels, you have strayed from empathy and listening. When you catch yourself straying in these ways, know that it is normal and inevitable, and then refocus yourself to feel what your partner feels. Seeking pure empathy in this exercise may frustrate you just like a meditation that focuses your mind on a single thought or mental behavior. Yet, in this case, rather than a single mental behavior, your task is to focus on a single emotional behavior. An irony is that once you realize you are on track with empathy, you are no longer on track. Pure empathy is purely feeling with the person communicating to you.

We estimate that you will need about 10 minutes of listening to your partner while you strive for empathy to learn well from this activity. Much more than 10 minutes may get too difficult and ask too much of the speaker. Less than 10 minutes will not give you enough practice time to contemplate. As usual, we suggest your partner be a classmate or other student of counseling and related fields.

After the session, journal and discuss what was difficult for you as you strove for empathy. Journal and discuss at what points your empathy was pure and at what points you strayed. Ask your partner if he or she can remember the points in communication in which he or she felt/thought your empathy was pure and when you had strayed. Ask the same of your observing partner. It may be helpful to review a video of the practice session together. When you review your video, check for moments when it seemed your empathy was pure and when you strayed. Then consider what you looked or felt like or what other indicators there might have been when your empathy was pure and when it strayed.

After the activity, journal and discuss what you have learned. Address what seems to get in the way of empathy for you. This could be certain subjects or emotions; it could be your urge to solve the other person's problem or tell about yourself; it could be unrelated thoughts that pop into your mind. After addressing what gets in the way for you, describe the signs (i.e., thoughts, feelings, physical sensations, facial expressions, posture, gestures) that you may be able to use to indicate to yourself when you have strayed from empathy. Additionally, describe the signs that let you know you are experiencing empathy.

INTRODUCTION

For something that occurs naturally, empathy can be very hard. We do not expect your work in the focus activity to have come easily. We see empathy as both so often difficult and so crucial that we have separated this chapter, "Striving for Empathy," from the next, "Expressing Empathy." We want you to spend more time in the study of this crucial skill. We do not want you to confuse the behaviors of expressing empathy with the skill of experiencing empathy and letting it be apparent in you.

Additionally, in our skill objectives we have indicated that you are beginning what may be your long and ongoing path to providing strong empathy. For us and for many we know, developing empathy and our ability to express it is a lifelong and continually rewarding path. From this chapter, we expect you to begin to understand the power of empathy, while fuller understanding will come from the experience of providing empathy in counseling and other relationships. We expect you to make significant progress in your ability to have and express empathy. We expect you to identify and begin overcoming factors that prevent you from experiencing fuller empathy. Once it is well begun, we expect you to experience this as a steady-flowing, ever-unfolding, and joyful endeavor.

WHAT EMPATHY IS AND ISN'T

Understanding What Empathy
Is by Considering What It Is Not

Empathy is not a thought process. If you try to think of what another person is feeling, you have taken yourself away from empathy, away from connection with the other person, and into the separate world of your thoughts. Babies know what others feel and react to it. If parents are upset, their baby will know and react, even if the parents keep themselves from crying, yelling, or otherwise showing their stress. In times like this, many babies will cry or react with emotion. For those who don't cry, if you watch closely, you can see them tense and become watchful and wary.

Symphonies evoke strong feelings in listeners without using words. Music scholars can often tell you what tools were used to convey these emotions. Yet listeners feel the emotions even without the intellectual skills to figure out what the composer may have meant to convey and how.

Jeff remembers a moment from childhood that helps him remember that empathy is not a thought process:

> My grandparents had a horse named Tony. I hadn't been around horses much, so I was a little afraid of him. Tony was gentle, but *so* big. We were going to saddle Tony when my grandfather told me not to be afraid around Tony because he would know it and then get skittish and jumpy. I did my best to talk myself out of my fear, reminding myself how gentle Tony was. I replaced some of my fear with remembering how much I loved Tony. Still, I was a little scared and after what my grandfather had told me, I watched Tony. I saw that he was a little skittish around me. I remember wondering, how could Tony know I was afraid, especially since I thought I was pretty good at hiding fear and couldn't see how my interactions with him were that different from my cousins, who were more used to horses.

If horses and babies feel with others and experience empathy, then empathy must be a natural occurrence and one that requires only basic presymbolic thought. Indeed, thought may be what gets in the way of or dilutes the purity and power of empathy. Our students sometimes ask us early in their work what they can do to develop their empathy and tell us that it does not seem natural to them. We see that their task is less one of developing their empathy skills and more one of dropping the things they do that get in the way of empathy. Empathy is natural to babies. As people grow up, their ability to reason develops. We often get so enamored with our ability to reason that we try to use it for most or all of life's tasks. It is as if we have discovered what a versatile and powerful tool a hammer is and we try to use it to replace a window—it works well for getting the old glass out but not for putting the new glass in. Thus, we stop developing our skills in empathy and sometimes even retard these skills. However, once persons begin to attend to their skills in empathy, these skills begin to grow again rapidly.

Empathy also is not sympathy. Sympathy is what we feel *for* others, rather than *with* them. When we know a friend has had a loved one die, we feel great sorrow and sadness for that friend. This is because we assume, often from similar personal losses, how badly our friend probably feels. Such sympathy can feel supportive and can be quite helpful, but it is not empathy. Persons experiencing such a loss feel sets of emotions unique to themselves and their situation, which might include sorrow, relief, anger, guilt, and other emotions. Both sympathy and empathy can convey caring, but they convey different messages of understanding. In this case, sympathy conveys, "I understand how badly it hurts to have a loved one die," while empathy can convey, "I understand you through what you feel right now."

Intricacies of Empathy

From our thoughts on what empathy is not, we see that it is deeply feeling what another person feels in a given moment in that person's unique life experience. Rogers (1980) described empathy as meaning that "the therapist senses accurately the feelings and personal meanings that the client is experiencing, and communicates this understanding to the client" (p. 116). Further, empathy means "to sense the client's private world as if it were your own but without ever losing the 'as if' quality" (Rogers, 1957, p. 99). We take the phrases "personal meaning" and "private world" to clarify that while empathy is deeply feeling with another person, it also indicates a totality of experiencing with the other person. So more broadly, empathy means fully experiencing another's world. The other person's emotions are key to entering her world, but empathy also includes her thoughts, beliefs, and perceptions; her actions; and the interplay between her emotions, thoughts, and actions. Empathy is experiencing the inner world of another, her full and unique personal experience.

The phrase "without ever losing the 'as if' quality" is also important, as empathy does not mean to lose your self in being only a mirror to others. Rather, empathy takes place between two separate persons while one is striving to experience with the other. Consider this longer quote from Rogers (1980) on empathy, which helps explain the delicate balance of remaining yourself in your counseling sessions while fully experiencing with your client:

> An empathic way of being with another person . . . means entering the private perceptual world of the other and becoming thoroughly at home in it. It involves being sensitive, moment by moment, to the changing felt meanings that flow in this other person, from the fear or rage or tenderness or confusion or whatever he or she is experiencing. It means temporarily living in the other's life, moving about in it delicately without making judgments—it means sensing meanings of which he or she is scarcely aware, but not trying to uncover totally unconscious feelings, since this would be too threatening. It includes communicating your

sensing of the person's world as you look with fresh and unfrightened eyes at elements of which he or she is fearful. . . . You are a confident companion to the person in his or her inner world. (p. 142)

Guerney (2002) refers to the level of empathy that we wish for you in your counseling and that Rogers described as Deep Empathy. Guerney explains that this shared experience is so strong that your empathic responses may sometimes include "the next emotion that the speaker might have said . . . the next wish or desire, or the conflict underlying what the person has told you, but did not verbalize" (p. 3). He explains that it is better to strive for Deep Empathy and risk being wrong than to maintain accurate but pedantic active listening (paraphrasing what your clients have said), with no risk of being wrong but also with no empathic connection.

Empathy is what makes therapeutic listening powerful. Empathy is what takes therapeutic listening beyond just the words, beyond mere paraphrasing. Therapeutic or reflective listening without empathy is a parody of counseling. It might be the outward behaviors of counseling but is still not counseling.

A SAMPLE OF THE PREPONDERANCE OF LITERATURE SUPPORTING AND CLARIFYING THE IMPORTANCE AND POWER OF EMPATHY

Rogers was the first to introduce the helping professions to empathy in a big way, introducing it as a key factor in the core conditions of counseling (including empathy, unconditional positive regard, and genuineness) (Rogers, 1957). We owe a great debt to Rogers and other thinkers, writers, and practitioners in the person-centered approach for this contribution and its ongoing development.

Research and professional literature in the helping professions has repeatedly confirmed the importance of the core conditions for therapeutic relationships (including empathy, unconditional positive regard, and genuineness). In their research review, Orlinsky and Howard (1978) concluded that there is substantial evidence in support of the relationship between clients' perception of the core conditions and outcome in counseling. Other authors reviewing research seem to concur (Bergin & Lambert, 1978; Krumboltz, Becker-Haven, & Burnett, 1979). More recently, Peschken and Johnson (1997) added that the core conditions are associated with clients' trust of their counselor. A large number of meta-analysis researchers assert that therapists' qualities, such as the core conditions, are crucial to and predominate explanations of counseling successes (e.g., Lambert & Okiishi, 1997; Luborsky et al., 1986; Wampold, 2001).

Patterson (1984) concluded in his review that "the evidence for the necessity ⁄ ... of the [core conditions] is incontrovertible" (p.437).

The core conditions, especially empathy, are valued across widely differing theories of counseling. Corey (1996) explained that most major counseling theories have incorporated the importance of therapists' ways of being in the therapeutic relationship as conducive to the use of their techniques. Corey offers the example that while cognitive behavioral therapy (CBT) offers a wide range of strategies designed to help clients with specific problems, these strategies are based on the assumption that a trusting and accepting therapeutic relationship is necessary for successful application. Keijsers, Schaap, and Hoogduin (2000) explained that the core conditions are clearly associated with successful outcome in CBT. Haaga, Rabois, and Brody (1999) concurred that CBT works best for therapists who convey empathy clearly to their clients. Lineham (1997) described empathy as a key element in dialectical behavior therapy; Kohut (1984) as a centerpiece in self-psychology; Scharf (1996), its importance in Adlerian therapy; and May (1989) and Havens (1986) from an existential perspective. O'Hara and Jordan (1997) simply concluded that there is no therapeutic relationship without an empathic therapist. Duan, Rose, and Kraatz (2002) summarized that because empathy is the basis for understanding, one can conclude that no effective intervention can take place without empathy, and all effective interventions must be empathic. In their meta-analysis, Bohart, Elliot, Greenberg, and Watson (2002) assert for the power of empathy in explaining positive outcomes and explain, "The time is ripe for the reexamination and rehabilitation of therapist empathy as a key change process in psychotherapy" (p. 89).

Recent literature seems to demonstrate a renewed and growing interest in the importance of empathy in counseling and also in a variety of noncounseling therapeutic relationships. This seems particularly important as counselors are called on to provide varied functions with differing relationship levels and purposes in today's school and other work settings. Additionally, these studies and essays can be particularly helpful in understanding empathy, its importance, and why and how it is therapeutic. Blake and Garner (2000) evidenced the importance of empathy in teachers of students with behavior disorders; Sweeney and Whitworth (2000) for the success of beginning teachers; Clifford (1999) in the connection between mentor empathy and protégé teacher self-efficacy. Its importance has been discussed in pastoral counseling, and within and between diverse faith communities (Augsburger, 1986; Everding & Huffaker, 1998; Kinast, 1984; McCarthy, 1992). Lenaghan (2000) asserted the importance of teaching students empathy for understanding cultural differences; Lickona (2000) for preventing peer cruelty among school children; Lo Bianco (1999) for training teachers of multiculturalism; Brooks and Goldstein (2001) for parents in developing children's resiliency; and Kountz (1998) for composition teachers assisting students in overcoming writing anxiety.

WHY EMPATHY
IS IMPORTANT, POWERFUL

Some explanations and examples of why empathy is important and powerful follow.

Connecting at the Core

Our emotions connect to our core selves. As empathy is an emotional, true, and honest understanding of one person from another, it allows clients to explore their deepest, darkest fears. At times in our lives, many of us seem to believe that if others really knew us, they would not like us. Perls (1970) describes personalities as layered with a phony outer layer that hides the true self, then a phobic layer of catastrophic fears, so that if others saw our true selves, they surely would reject us. In other writing, we have explained how children with conduct-disordered behaviors (CD) often seem to believe that at their core they are unlovable and unlikable, so they act out to drive others away from understanding this core (Cochran & Cochran, 1999). In the classic film series in which Gloria serves as client for three different therapists demonstrating their theoretical approaches, she seems to tell her therapist that if he really knew her, he would find her unlovable (Shostrom, 1965). So by providing empathy, we help our clients show us their true selves. Then, in expressing empathy, we help them recognize what it is that we see and experience. In the case of our conceptualization of children with CD, your clients may come to see that you really came to know them yet were not repulsed by what you saw and experienced.

When experiencing your empathic understanding of their core, some clients then decide they truly do not approve of what they see and hear from their therapist's expression of empathy and so make changes in themselves. Other clients decide that the rotten core they feared is really not so bad, that they are lovable as they are. When it happens this way, we think of sailors from before the time people knew most of what is within the dark ocean. When humans didn't know what was in the water, many imagined great monsters and felt great fear. Now that we know more, we know that there are things to fear, but there probably are no great sea serpents or other monsters. Another metaphor that comes to mind is a child who sees scary shadows in his dark room at night and imagines unseen dangers. Then, when a trusted adult helps him turn the light on, he discovers that it is only a coat over a chair, or when searching under the bed with a flashlight, he learns there was no creature ready to grab him, just old dust and lint. A counselor's empathy can be like turning on the light to see what there really is to fear in our dark shadows.

Joining on a Scary Journey

It is a scary thing to explore our hidden emotions and the deep recesses of our selves. Many of us heavily defend the secrecy of these parts of ourselves. Like sailing with our modern knowledge of oceans or spending first nights as a child in dark rooms with shifting shadows and imagined creatures under the bed, once it's been done a few times, we know we are all right and our fear diminishes. But part of what's so hard about these first few times is going into the scary unknown alone. Your empathy communicates to your clients, "I'm right here with you," while they search into their scary places and their deepest fears. We've had experience in accompanying children on outdoor and water adventures. In those experiences, the children's autonomy and self-responsibility grew, but it helped the children to meet their challenges for us to first communicate with our actions and even in our words, "I'm right here with you." As Rogers (1980) wrote, "You are a confident companion to the person in his or her inner world" (p. 142). Your empathic presence is your message, "I'm right here with you," that gives your clients the strength to look into the shadows. You, with your empathy, become your clients' opportunity to examine those parts of themselves that they might never explore alone.

It is important to add here that while you may share these explanations of the power of empathy with your clients, usually you won't. Most clients do this work without ever completely articulating what they and you are actually doing in their counseling. We add this note because explanations and examples of counseling are sometimes like highlight films of ball games. The highlights can make it seem like the whole game was ecstatically dramatic. Yet, viewing the whole game would show a lot of action that both appears mundane and was really hard work as players struggled to maintain their concentration. We want you not to feel discouraged when your and your clients' experience is long, hard work together, rather than just highlight film.

Self-Awareness

As you join your clients through empathy, they come to further know themselves, how they work, and why. Rogers (1980) explained, "As persons are empathically heard, it becomes possible for them to listen more accurately to the flow of their inner experiencing" (p. 116). We also think there is a modeling effect. As you, their counselor, attend to their inner experiencing through empathy, your clients will attend to themselves similarly.

Self-Experience

Through your empathy, clients come to experience their emotions more fully and deeply. This fuller experience is quite durable once started. Whether during or after sessions, it is often the level of emotions that clients feel that tend to prompt action. Often, a person's level of experienced discomfort with a current situation needs to become greater than her fear of change in order to produce change. At other times, this fuller self-experience helps persons realize that the answers to dilemmas they have been unsuccessfully trying to solve through a logical analysis rest in their emotional response. For example, decisions for how to respond to others in relationships, whether to continue or begin a romance, often cannot be reasoned from the mind but can be decided based on a full understanding of what is in the heart.

A Profoundly Different Relationship

Another power of deep and full empathy is that it is so rare. It is often the profound uniqueness and depth of an empathic counseling relationship that makes it powerful. As a child with conduct-disordered behaviors goes through life, he experiences the relationships that he expects. That child may expect to be disliked and so drives others away to avoid the pain of that rejection. In response to these behaviors, others do recoil from him. Yet, in an empathic counseling relationship you respond very differently. You respond to that child's feelings much more than to his actions. Thus, you see that child not as a set of repulsive behaviors but as a person with feelings. Such a child will then experience a useful confusion. While that child's old self-statement sounded something like, "No one can love me. Everyone will reject me," your relationship begins to change that statement to one that sounds something like, "Well, I thought everyone would reject me, but my counselor didn't, and she really knew me." Such clauses in negative self-statements are the beginnings of great change in the emotions and behaviors driven by such statements.

Some children and adults enter counseling expecting to be told what to do, how to act, and how to be. Often, this is because such telling, whether subtle or blatant, permeates and defines the relationships they have with others in their life. If you don't offer them this same direction, it may present them with an anxiety-producing experience. If, in their other relationships, they are rescued from responsibility with direction, they will expect and want this from you. If you meet them with an empathic acceptance of the anxiety they experience, experiencing it fully with them rather than working to make it go away, they also experience a useful confusion and must face the challenge of reconciling your reaction to them and their anxiety with the very different experience they have had with others. Here, many clients seem to shift from a belief that says, "I am not capable and need others to decide for me," to one that says, "But my counselor, who seemed to really know me, didn't seem to believe I needed her to decide for me; she experienced my anxiety and didn't think we must work to make it go away." So many of the clients we have served have found this experience greatly empowering, although awkward at first. In a side note, some adolescents and adults may also need you to help them understand what the two of you can do in counseling and why they need to tolerate and experience this anxiety in order to work through it. We illustrate such situations and useful reactions with explanations in Chapter 8.

Joy in Connecting

We'd like you to know that the power of empathy isn't just about pain. Many clients experience great joy in their counselor's empathy, as they realize their counselor "gets them." We have experienced a great many clients who took to empathy from their counselor like a thirsty person takes to water on a hot day. Jeff recalls one such client:

A young man that I served had great difficulty fitting in socially. His interests were certainly not with the mainstream. Plus, he often seemed to say and do the wrong thing socially. He very slowly came to explain that he knew he didn't fit in, that he hurt with loneliness, and that he feared he was missing out. He was only direct with such communications in furtive, brief moments. Most of his time in counseling was spent telling me of the things that interested him very much but that most people would not take the time to listen to. As I experienced excitement with the things that excited him, not necessarily because they excited me but because I felt his excitement, he seemed to come to experience joy and pride that someone "gets him," not always his interests, but him. His self-talk seemed to change from, "No one will understand me," to "Others may not understand my interest, but they may come to understand me." From this change, he moved on to a successful group counseling experience (a process group, in which persons come to interact and share themselves, rather than a psychoeducational skill training group) and a general change in reaching out and risking letting others know him.

In Nancy's work with parents and children, she sometimes trains parents in filial therapy (Guerney, 1964; Guerney, 1976), which includes teaching parents to play with their children with a strong quality of empathy and acceptance. As children involve their parents in their play, they are given a chance to express the inner workings of their worlds, minds, and hearts freely and without fear of correction or punishment. The joy in experiencing this connectedness with their parents becomes powerfully clear and can be a beautiful process to watch unfold. The look of the parents when the deep connection is working is sometimes described as "the filial glow." This "glow" is always evident when relationships are enhanced through deep connectedness. Seeing the parent–child connection during filial therapy helps us see that all relationships are opportunities to experience this "glow"—this connection of caring.

Furthering Communication and Connection

Providing empathy has the effect of clients wanting to tell you more and share more of themselves with you. As we think of ourselves receiving empathy, we know that we have felt warmly toward and satisfied with others who have come to understand us with empathy. We've felt motivated to then let them know more of our story, kind of like a storyteller who checks the audience's eyes for interest before deciding to cut a story short or let it grow into the next, bigger, and more revealing story. So we see that empathy lends momentum to counseling. Still, getting your client to tell you more is only a side effect, not the real point of empathy. The real point is for your clients to experience a connection with you in which they would like you to know more and more about them. The connection is worth more than the telling.

WHAT GETS IN THE WAY OF EMPATHY

To experience empathy, you need to open your self to others' feelings and to your own. This connection works at least two ways. As you come to feel more fully and be okay with that, you will become more ready to feel the emotions of others and to let what you feel with them be apparent in you. Further, your tolerance for emotions and their ambiguities and intensities will increase. Additionally, experiencing emotions with others will elicit emotions in you that will include both the emotions of your clients and those unique to you. You may remember similar experiences in your life and feel your own hurt that is unique to you and not of your clients. You may fear for the further pain that your clients seem destined to experience, and you may worry for their safety. Experiences like that are not empathy but are part of the closeness that empathy brings.

Considering the need for you to develop your ability to feel your emotions and those of others more and more deeply, we ask you to search the following barriers that can get in the way of empathy for what gets in the way of *your* empathy and how so. We find that many of these barriers have or still do affect us, and since we like and are pleased with ourselves by this point in life, we certainly see no shame in identifying with these barriers. We challenge you to discover any other barriers to your empathy. We will be glad to hear what you have discovered, if you choose to write and tell us.

Habit

First, painful feelings tend to hurt, so we avoid them. It seems that in American society we have developed many ways to avoid feelings. We have TV and movies that take our minds off upsetting personal situations, drugs and alcohol that we use in an attempt to obliterate painful emotions, posters of cute puppies that tell us to cheer up, and an increasingly common habit of keeping ourselves too busy to feel anything except the constant urge to hurry onto the next thing.

People often get out of the habit of empathy and experiencing feelings just from becoming enamored with our ability to reason. For many of us, the ability to reason comes to seem a skill for all occasions. However, just as in life there are times when experiencing strong emotions may actually be problematic and limiting, overuse of our ability to reason is problematic and limiting in counseling. Reason is a useful tool, but like any tool, it is useful in some jobs and not in others. For a metaphor, money can fix many problems in life. If I need a car to get to work, money can help. But if I were lonely and wanted to find a spouse to truly love and be loved by, money could only help in peripheral ways, if at all.

Fear of Feelings

Beyond the basic avoidance of what hurts, many of us seem to become afraid to feel as we grow up. Some of this fear appears to be social programming from hearing such statements when emotional as "Big boys don't cry," "Don't be such a crybaby," or in some harsher instances, "Dry it up or I'll give you something to cry about." Here, the fear of feeling seems to be a self-statement that sounds something like, "If I show strong emotions, I will actually show weakness and if others see this, I will feel ashamed, embarrassed, and that will be awful." Jeff relates a part of his experience with this:

> I have learned that for a time in life I chose to feel anger instead of other emotions. I didn't cry. When others did, it was very upsetting to me. When others cried, it seemed like an internal panic started in which sirens sounded, lights flashed, and a voice from a loudspeaker extorted me, "Do something! Can't you see they're crying?!" Now, with much work, conscious effort, and help from loved ones and counselors, I have come to know that I can cry, even in front of others, and it's okay. Now, when others cry in my presence, I sometimes see just one flash of red light in my mind. I know it is the old alarm starting. I remind myself that the important thing is to be fully present with this person, which includes feeling what she or he feels. I sometimes also remind myself that this person, whether a loved one or client can be okay feeling the feelings that produce her or his tears, and so will I.

We have known some persons to fear strong emotions and to recognize that their fear is based in a thought that once they begin feeling painful emotions, it will be like opening a floodgate; they will be overwhelmed and swept away. Many clients have told us thoughts similar to this, then used counseling as a safe place, a place where they are accepted but autonomous, on their own but not alone; a place to try letting themselves feel strong painful emotions and see what happens. For some of these clients, it was their diligent defense against painful emotions that kept them from connecting with others, kept them from feeling joyful emotions, and kept them overly stressed and not well.

For many counselors, just like for some of the clients mentioned in our earlier section, "Why Empathy is Important, Powerful," the fear is that our core is rotten and if we and others could see it, it would be known, and that would be a terrible, terrible thing. So, with a mindset like that, a counselor can be prohibited from feeling strong emotions with clients. Such a counselor can get a sense that emotions lead to our core, may even *be* our core. So, the empathy-prohibiting belief of this counselor could be that "Certainly, I must not feel." And the application of that belief would be, "So, instead of working to feel with clients, I must work to prevent and squelch strong emotions."

Misattributed Responsibility

Sometimes counselors mistakenly see themselves as responsible for a client's pain or at least responsible for making it go away. Since empathy leads to more and deeper emotions, it can seem like it causes them. But *cause* is the wrong word for it. A more fitting description would be that empathy *allows* more and deeper emotions, that it provides a safe place and time, with a listening ear.

Additionally, some counselors worry about errors in their empathy. Because empathy is assisted by but does not rely on emotions being described in words, there is the risk of being wrong, of perhaps going off on a tangent of feeling that is misshaped by your personal feelings that are separate from your client's. Some counselors worry that this would produce a new feeling in their client. We encourage you to strive for accurate empathy but to avoid letting concern over error inhibit your empathy. Your empathy, even if wrongly expressed, will not produce significant new emotions in your clients as long as your intent is true empathy. Following our guidance in Chapter 2 on therapeutic listening and in Chapter 4 on expressing empathy will also help you avoid making errors that prompt a new feeling, beyond a moment of confusion from your client, as he realizes that you misunderstood him, or a momentary pique at being misunderstood.

Rather than something to worry over, when a new feeling does seem to develop in your client from your response, the new feeling simply becomes your client's next communication to you. For example, if you shed a tear of empathy with your client, your client may react by stopping her crying, feeling concern and guilt at thinking her situation has hurt you. When we have seen counselors respond with empathy to their client's concern and guilt, their new empathic response has proved a valuable learning experience to the client. In one such situation, the client learned just how much she tried to protect others, even her counselor, from feeling for her. She then experienced a strengthened motivation to be and feel who she is and to allow the same for others while being right there with them in their pain.

It is not the responsibility or even the purpose of the counselor to make feelings go away. Ironically, while persons often come to counseling to make painful feelings go away, and while that normally happens through the process of counseling, suppressing feelings cannot be the purpose of counseling or it will greatly limit the potential of counseling. Emotions are useful to experience (i.e., discovering that you are lovable, not rotten; learning who you really are and what you want; making fully self-informed decisions) and can be destructive to suppress (i.e., remaining partially blind to your behaviors, their meanings, and relation to self; experiencing somatic pain from working so hard not to feel or show feelings). If you as counselor want too much to make your clients' painful feelings go away and are unable to be okay with what they feel, your empathy will be inhibited and your clients' suffering prolonged.

This is not to say that we would never teach a client the skills for coping with painful emotions, which may be aimed at reducing them, when this is clearly what the client asks for. Rather, before doing so, we believe it is best to help the client understand what it is she is asking for, what alternative uses she can make of counseling, and the ramifications of taking such an approach in counseling. Reviewing Chapter 8, in which we provide scenarios, examples, and discussion of explaining use of counseling, will help with such situations.

Letting Go of Control

Empathy requires partially letting go of control over what you feel. This can be scary to counselors as it can threaten your believed/perceived need to be in control and stimulate fears of feeling. As your feelings follow your client's, you can reassert control at any moment by pulling back a little and becoming thoughtful. To be most efficient, your ideal should be to pull back in this way as little as possible. Rogers (1980) addressed this challenge to counselors:

> To be with another in this way [with empathy] means that for the time being, you lay aside your own views and values in order to enter another's world without prejudice. In some sense it means that you lay aside yourself; this can only be done by persons who are secure enough in themselves that they know they will not get lost in what may turn out to be the strange or bizarre world of another, and that they can comfortably return to their own world when they wish. (p. 143)

We wish for you to work hard and steady at this crucial skill in order to be maximally helpful to a great many others. Our hopes are with you on this worthy path.

ACTIVITIES AND RESOURCES
FOR FURTHER STUDY

- First, note that we have provided a great many activities and resources for further study for this chapter. Experiencing empathy is crucially important to the success of your work and your ongoing development. Please make time for as many of these activities as possible.

- Repeat the focus activity at the beginning of the chapter now that you have come to a greater understanding of empathy and potential barriers to empathy for you. We encourage you to repeat the activity as often as possible until you are satisfied with your consistency and learning. We especially urge you to take time after each practice to contemplate and discover the barriers to empathy you experienced.

- This same exercise and others like it in this book present a great opportunity, especially in the role of speaker, to develop your strength in feeling what you feel, learning what feelings come when you strive to express your feelings, or exploring through your communication your barriers to feeling your feelings and to empathy.

- We urge you to spend significant time contemplating, journaling, and discussing your barriers to empathy and how you may overcome or lessen them.

- We highly recommend you read Myers, S. (2000). Empathic listening: Reports on the experience of being heard. *Journal of Humanistic Psychology, 40,* 148–173.

    This qualitative research supports an in-depth understanding of what it means to really listen, the relationship between empathy and listening, and how effective therapists communicate that they are listening and understanding. Myers accomplishes this through research interviews that help clients articulate how powerful and important their experience of being heard has been and why. While this research is helpful for experienced therapists, it is particularly helpful in addressing some of the fears, inhibitions, and misunderstandings of beginning therapists.

- We have known beginning counselor supervisees to benefit from watching their session or practice session tapes while stopping themselves from evaluating their work and working solely to feel each client's emotions. This has helped them be ready both for the emotions that are thematic for that client and, in developing their empathy skills, to be more ready for all clients. You may also follow this procedure with your tapes of practice sessions with peers/classmates.

- A variation is to review session tapes with the sound off with the sole purpose of feeling with the client in order to become less dependent on words.

- Another variation is to watch a well-acted movie segment, with or without the sound, with the sole purpose of practicing empathy, feeling what the characters feel.

- The same could be done while listening to music, especially music without words and music that evokes a range of emotions, like many forms of classical music.

- To remember that you need not name the emotions (a cognitive, rather than empathic activity) in these exercises, you may want to practice or read about a form of meditation called *contemplation,* or "one-pointing" (LeShan, 1974, p.73). This approach is not the same thing as empathy, but it can help you master skills required for excellent empathy. In this approach, you are to contemplate an object without using words. It can be difficult, as we humans are often so enamored of words and the naming of things, and it can be eye opening. We like LeShan's book, *How to Meditate: A Guide to Self-Discovery,* for its concise, readable, and practical explanations and guidance.

- We strongly urge you to carefully read works by Carl Rogers and other writers and thinkers in the person-centered approach.

- We encourage you to review works by the National Institute for Relationship Enhancement, as persons involved with that institute have done great work in operationalizing empathy in counseling, family, and other relationships.

- We encourage you to take a look at the *Person-Centered Journal,* as it often prints articles discussing topics involving and related to empathy.

- Because empathy is not the exclusive property of persons with interest in the person-centered approach and is of interest to persons across the helping professions, we also urge you to seek out the many discussions of empathy and its importance in various conferences, journals, and groups of friends and colleagues in the helping professions.

- We of course refer you to the sources cited in this chapter and listed among our references.

- Take time to contemplate the main ideas of this chapter and your reactions to them. Journal and discuss your reactions.

- Review the Primary Skill Objectives of this chapter and see that you have mastered each to your satisfaction, for the moment. If you have not yet mastered them to your satisfaction for the moment, spend more time rereading and contemplating the chapter, and more time in practice, and then carefully note the areas in which you are dissatisfied to keep in mind for opportunities for continued improvement while studying upcoming chapters.

- And last but certainly not least in our values, we suggest you seek a strongly empathic counselor for yourself, especially to give yourself the time to work through the barriers you have to empathy and feeling fully.

4

Expressing Empathy

Every vital development in language is a development of feeling as well.

T. S. ELIOT

PRIMARY SKILL OBJECTIVES

- Be able to explain the importance of expressing empathy.
- Understand and begin to successfully implement the ways to express empathy discussed in this chapter.
- Understand and be able to explain the Do's and Don'ts of expressing empathy.
- Successfully implement the Do's and Don'ts of expressing empathy.
- Understand and begin to successfully implement the nuances of expressing empathy discussed in this chapter.
- Understand and anticipate the effects on your work from the common difficulties, pitfalls, and dead ends in expressing empathy that are discussed in this chapter.
- Evaluate and be able to explain your strengths and difficulties in expressing empathy at this time, how you may overcome your difficulties and

develop your strengths, and what you expect your ongoing strengths and difficulties to be.

- Be able to explain what expressing empathy looks, sounds, and feels like to you when you are working effectively.

FOCUS ACTIVITY

Now it's time to let yourself express the empathy you feel. Have a partner communicate to you about a topic that has emotional content for her or him. Strive to feel what your speaker feels and to express what it is that you feel with her or him. Your expression of empathy may take a variety of forms. Try not to worry about the form your expression of empathy takes. This worry would distract you from empathy and listening. For this exercise, you are adding one intention beyond experiencing empathy, which is expressing the empathy that you feel. An irony to work through will be that while you aim to express your empathy, you can't think too much about how to express your empathy because that will take your focus off empathy with your speaker. So, a useful way to think of it may be to think of removing the restraints that we asked of you in the focus activity of the previous chapter and freeing yourself to express what you feel with your partner.

You should also practice therapeutic listening with this activity. It would be hard not to, as much empathy can be expressed in the tone of your reflections. But don't let yourself worry right now about things like reflecting. Rather, focus on freeing yourself to experience and express empathy.

A reasonable length of time for this practice is 15–20 minutes. Have a third partner observe and/or video tape the practice. Use feedback from the speaker, observer, and video review to discern when your empathy was clearly expressed and when it was not. Also, use their feedback to discern when you had apparently strayed from empathy and therapeutic listening.

Please note that when you are in the role of speaker, you may be able to give your listener helpful feedback regarding her or his listening, empathy, and expression of empathy. However, you really can't focus on watching for your listener's skills while speaking. Rather, it will be best if you purely focus on expressing whatever comes to your heart and mind. Then later, if you remember something of your listener's skills it will be helpful and nice to tell her or him, but the most help you can be to your listener and yourself is to maintain your focus on you and what you wish to say.

Discuss with your partners, a small group, or with your class, and/or journal what you have learned from this exercise. Describe the ways of expressing empathy that seemed to work for you and the ways of expressing empathy that you hear or see in others' work that may work for you. Remember that empathy is not sympathy or thinking of what others feel. Notice and describe when sympathy and thinking may have gotten in the way of expressing empathy. Also describe any other challenges you have encountered and antici-pate in expressing empathy.

One final reminder with this exercise: Do not be discouraged if this work is difficult at first. If it is difficult for you, you are in good company. If you are work-ing carefully and finding that it comes easily, enjoy, as that is not always the case.

VARIOUS WAYS TO EXPRESS EMPATHY

Probably the purest way to express empathy is through the emotion that is returned to your clients from you as you feel with them—the emotion they can feel from you as you feel with them. As you may have found in the prac-tice activities for this and the previous chapter, overt expressions of empathy do occur when striving only to feel, without even an attempt to express empa-thy. Your work will be best, you will be most focused, and the process will be most natural when you are focused on feeling and allowing yourself to express empathy versus being focused on your expression.

Ways of expressing of empathy cross an array of modes, from involuntarily emitting the emotions that your clients simply feel from you as you feel with them, to saying the words for what you feel with your clients. We don't see any one method as necessarily better than others. Which is best may be determined by which comes naturally to you in each unique situation and can be determined by trying what feels most natural to you in each moment, then observing if your clients seem to perceive your empathy (or asking if and how after practice sessions). Next, we introduce a few ways of express-ing empathy. Each can be a natural outgrowth of your striving to feel with your clients.

Matching Client's Tone

As you feel empathy, your tone of expression will match your client's. Your voice will tend to rise and fall following your client's. Your pace of speech will tend to quicken and slow following your client's. In watching a tape of your session, you will be able to hear the quality of your client's emotion in your spoken responses.

Facial Expression and Body Language

We discovered some time ago that neither of us has a "poker face." Not having a poker face is actually an asset to us as counselors. What we are feeling can be obvious to our clients or to anyone watching. So, when we strive for empathy and succeed in feeling with clients, our empathy is expressed through our faces mirroring the emotions of our clients.

We have watched sessions by counselors under our supervision in which their empathy was clearly expressed through hand gestures and other body language. For example, one client was explaining how her father would sound and move when he was about to angrily come after her and her siblings. When she made the sound and threatening move, both she and her counselor made a body gesture that suggested ducking one shoulder and head while starting to run. That gesture symbolized the fear-filled response that she and her siblings had. The counselor was not mimicking her client. It was a gesture she made while absorbed in listening and empathy, without thought of expressing empathy. Her client did not comment on the gesture but clearly knew in that moment that her counselor was with her.

In other moments, we have watched counselors clasp their hands to express empathy when a client expressed a feeling of connectedness, or shift back and move their hands wide apart when their clients excitedly expressed that what they are saying is *really* big for them. We have seen counselors push their fists together and grimace when their client expressed dread over an impending conflict. We have seen counselors brush their hands against each other as if dusting off sand or fling them out lightly as if flinging the last drops of water off their hands into the sink when their clients expressed that they were done with a troubling situation. Importantly, each of these gestures was natural, not contrived; not thought of, but born of feeling with clients and allowing expression of empathy to flow.

The Most Overt Means—Words

Sometimes the gestures, facial expressions, and tones that served to express empathy were also accompanied by words that named the feeling. Naming the feeling in words is what most people think of when thinking of expressing empathy. Using words to name feelings can make your empathy most overt. It can make what you feel in empathy least subject to interpretation, at least for persons of similar linguistic and cultural backgrounds.

However, expressing empathy through feeling words also brings its problems. When you work at thinking of the word for the emotions, then those thoughts take you away from empathy and listening.

Additionally, if you work hard at thinking of the word, your response often comes out sounding like an assessment, or the answer on a quiz show. When this happens, your client will tend to stop expressing and stop feeling. In these moments, it feels to clients as if that was somehow the end of the conversation or that you have made your diagnosis, so they wait for your prescription. It will help you avoid this to remember that your task is not to identify feelings so much as it is to help your clients feel more fully and to be right there with them through it all. Your saying the word for the feeling helps your clients know that you are right there with them, that you are okay with what they feel, not afraid to feel what they feel and go with them to the emotional places they must go, that you are "a confident companion . . . in his or her inner world" (Rogers, 1980, p. 142).

DO'S AND DON'TS
OF EXPRESSING EMPATHY

Empathy can be an esoteric concept. It is abstract, complex, and subtle. Yet, we want to make it as concrete and observable as possible. So, with our previous explanations of the purpose of empathy and some varied ways in mind of expressing it, consider the following "Do's" and "Don'ts," which allow us further discussion and which you may use to check your work in expressing empathy.

Overview

You may think of this Do's and Don'ts list as an extension of the Do's and Don'ts list for therapeutic listening. Experiencing and expressing empathy involve listening at an even deeper level. Experiencing and expressing empathy in your meetings with clients are in addition to therapeutic listening. The balance of therapeutic listening with experiencing and expressing empathy will differ for each client and situation, but you will always be balancing responding to communicated content with therapeutic listening and responding to experience by expressing empathy.

With these behaviors, you are communicating truths like, "I understand what you feel and experience it with you," "I understand your situation," "I sense what is important to you," "I'm striving to feel as much as I can with you, to feel as if I were you," and, ultimately, "Through experiencing with you, I'm coming to understand you."

Do's

___ Focus your attention primarily on client emotions, secondarily on thoughts and actions.

___ Strive to feel with your client, to feel what your client feels.

___ Reflect client feelings and underlying thoughts through the tone, facial expressions, body language, and gestures that come naturally to you.

___ Reflect client feelings with words for the emotions you feel with them, when natural.

___ Reflect feelings and underlying thoughts that you perceive your client to imply.

___ State your empathy in declarative statements, when reasonably sure.

___ When unsure, state your empathy as tentative, with more tentative declarations from your struggle to understand your client's feelings and underlying thoughts.

___ Use reflections to restate client feelings and underlying thoughts more clearly and directly.

___ Use reflections to restate client feelings and underlying thoughts more precisely and concisely.

___ Be prepared for and accept corrections of your empathy.

Don'ts

__ Don't let your words for what you feel with clients sound like assessments.

__ Don't respond with a hidden agenda of what you believe clients should realize.

__ Don't do most of the talking.

__ Don't make "me too" or "must feel" statements.

EXPLANATIONS AND DISCUSSION FROM THIS DO'S AND DON'TS LIST

Much as we may try, the complex and subtle skill of expressing empathy can't be truly captured in a Do's and Don'ts list. So, please carefully consider the following explanations and discussion.

Focus Your Attention Primarily on Client Emotions, Secondarily on Thoughts and Actions

Full empathy is a total connection with a person in which you understand the person in ways that go beyond the information the person tells you about himself to experiencing that person's world through his communication to you. Sometimes your expressions of empathy may reflect thought patterns and actions. For example, "[Stated with a tone and gestures of anger, agitation, excitement with client] You decided he'd crossed a line and you yelled at him for it!" Expressions of empathy can address both emotion and action, "You're so mad about this you can hardly sit still." At other times, we have seen naturally occurring expressions of empathy focus on a client's implied thoughts. A client was expressing his exasperation and amazement with coworker errors and his critical judgment of this. His counselor responded with the same exasperation and amazement in her tone, "So, you're thinking like, 'What are they thinking?! They must not be thinking!'" Still, while we hope for you to free yourself enough to respond with empathy to clients' thoughts and actions, we want you to keep this secondary to your focus on and expression of empathy to client emotions. *Emotions lead most efficiently to a connection with each client's core and experiencing each client's world as that person does.*

Strive to Feel with Your Client, to Feel What Your Client Feels

Remember that this is almost always your primary purpose and function. Don't get wrapped up in things like analytic thoughts, thoughts of what you'll say next, or concern for your client's satisfaction with your work. These things can be parts of your work (we will address them later), but the main part is feeling with your client. This is the core of your work.

Reflect Client Feelings and Underlying Thoughts
through the Tone, Facial Expressions, Body Language,
and Gestures That Come Naturally to You

Remember that these modes of expressing empathy can be as powerful and clear as using words to express your empathy. Sometimes these ways of expressing empathy catch more of clients' subtle communication. In one session, a client was expressing her dread and frustration at her mom taking a former boyfriend, who had been abusive in the past, back into her life. The client was wanting to be okay with it and supportive of her mom but already saw her mother changing to allow him back, even changing house rules to fit his needs in ways that she would not have changed them for her daughter. When the client said this, her counselor uttered a sound that was somewhere between a groan and a growl. His client heard this, and while she didn't stop to acknowledge it, she glanced into his eyes, grinned, and picked up her pace and the emotional intensity of her expression. She knew that her counselor felt her feelings that went along with both her dread of what she expected to follow from the situation in her family and her wish to drive the ex-boyfriend off and protect her mom, her territory, and herself. Such unplanned expressions of empathy happen naturally, when you are feeling with your client and allowing your expression to flow.

Reflect Client Feelings with Words for the Emotions
You Feel with Them, When Natural

Use words to express the emotions you feel with your clients. These can be feeling words, "You feel so mad about this!," "Now you are really hurting," or they can take other forms, such as expressing client thoughts and actions (as in the previous examples) or expressing empathy through metaphor, "It's like you're slogging hard, through cold rain and snow, just to get through your day."

Reflect Feelings and Underlying Thoughts
That You Perceive Your Client to Imply

Our choice of the words "that you perceive" is important to us. We don't want you to be so overly concerned with accuracy in your empathy that it becomes inhibiting or only allows you to respond to the emotions that your client has already named. You will sometimes reflect emotions your client has clearly stated, and you often will reflect the ones that it seems you feel with your client but that your client has not stated. Just as with therapeutic listening to the content of what your client communicates to you, *there is no penalty for being wrong in expressing empathy*. If you have misunderstood, your client can correct you and in correcting you, reach a greater self-understanding and fuller experience.

State Your Empathy in Declarative Statements
and, When Unsure, State Your Empathy
with More Tentative Declarations

When your client's emotion is strong and you are sure of what you feel with her, state it strongly. For example, "You're tired of hurting with this. Now you're mad and deciding not to take it anymore!" If you are less sure, you may make a more tentative statement, "Sounds like you feel confused and unsure of what to do next." If you are greatly unsure and think you may be too far out ahead of your client but still think it's important to say, you can express your experience in an even more tentative phrasing, "I'm not sure, I know you're exasperated with him, but as you talk, I get the feeling you are also very hurt." Each of these statements attempts to say what the counselor feels with her client. They also say, through behavior, that the counselor is striving or struggling to understand her client, and that while the counselor would like to be accurate in empathy, she is willing to risk being wrong in order to understand deeply.

Please note that no matter how tentative, these expressions of empathy remain declarative statements. They do not include a questioning tone. We find that asking a client what he feels almost always elicits a thought response (i.e., an explanation of feeling or what the client thinks he is supposed to feel), rather than helping that client feel more of his true feelings with you.

Use Reflections to Restate Client Feelings
and Underlying Thoughts More Clearly, Directly,
and More Precisely and Concisely

As your clients strive to express themselves to you and you strive for empathy, it will often be clear to you just what they feel or think when it is not yet clear to them. Because you are separate from their feelings, you may experience the feelings with less inhibition, fear of feelings, and confusion. For example, a client who has been hurt by a loved one might have a hard time expressing herself. She may try to communicate to you but struggle to get words out that express her feelings. Her statements might come out something like, "[Pausing between phrases and speaking in spurts] I just can't believe he did that . . . It makes no sense . . . I'm so mad . . . , but how could he? Didn't he think of me at all?" Hearing your client's tone and knowing the context of what she is saying, you may not only reason that your client is feeling shocked and hurt, but even more decisively and through emotional connection to your client, you may clearly *feel* the shock and hurt as she speaks. Thus, you may respond tenderly, "As you see what he's done, and tell me about it now, you feel shocked and hurt." It will be easier for you, as you strive to feel with clients but remain separate, to experience and state feelings more clearly, directly, precisely, and concisely. Through listening and feeling with them, you will be able to get to the heart of your clients' communication to you more quickly and easily than they can. This is one of the benefits you offer your clients.

Be Prepared for and Accept Corrections

Being wrong in your struggle to express empathy is not a problem and can even be beneficial, as long as you truly are striving to feel and express empathy. When a client corrects us after making an error in empathic understanding, we usually respond with something like, "Oh, I misunderstood there. You are very angry but not hurt. It is important to you that something like this cannot hurt you."

Sometimes corrections are not quite overt and only come in a feeling or facial expression. For example, after reflecting anger with a strong tone, if my client crinkles his face and lowers his head, and I get a feeling like the wind went out of his sails, I may reflect, "Something happened when I responded to what I thought was the strength of your anger. Seems like you suddenly lost your momentum."

In both of these examples, the client's response to the counselor's error and the counselor's acceptance of this response helped the client further realize his experience and reiterated the importance that his counselor places on truly understanding and experiencing with him. It is the same as being corrected in therapeutic listening to content. There is no point in a counselor going back and explaining why she thought her expression of empathy was true. By the time the correction or client response to the counselor's statement has occurred, the client has moved on to a new expression.

An exception to not going back and explaining might be when it seems clear that a pattern of misunderstanding persists. For example, if you believe your client has repeatedly expressed emotions that are similar to anger, and you expressed your empathy with these angry emotions in your tone and words, but your client continually rejects your expressions of empathy, it may be important to comment on the process of interactions between you. This comment may be something like, "I'm trying, but I seem to keep misunderstanding you on this. Several times I've felt sure you were really mad about this situation, but you've let me know that I misunderstand you." Such responses can open the door for your clients to express what they experience in relationship with you or to further clarify for you and themselves what it is that they are experiencing.

Don't Let Your Words for What You Feel with Clients Come Out Sounding Like Assessments

Remember that using feeling words can be a useful shorthand. Feeling words can make your empathy most overt, most quickly. However, using them also carries pitfalls and difficulties. Remember when using feeling words that your point is expressing what you are feeling with your client, rather than finding the correct word for what your client feels. Finding the correct word is a cognitive, problem-solving function. Work to minimize such assessment functions in your counseling relationships. Such assessment functions have the effect of inhibiting useful client experience and expression, and tend to be disempowering.

Don't Respond from a Hidden Agenda of What
You Believe Clients Should Realize

While we encourage you to risk being wrong in expressing empathy and to know that even errors in empathy can be beneficial, it is important to check that you are not wrong because you have an agenda other than understanding your client. For example, some beginning counselors may see clients' lives or dilemmas as simple, when they are not. This mistake seems to come from a cognitive error like the following: The counselor cares for his client in such a way that he almost believes that he can't stand to see his client suffering more. Thus, the counselor convinces himself that one emotional realization or another will bring a swift and sudden relief of suffering for his client. However, this is seldom true as life in and out of counseling is rarely that simple. This is an important error to avoid, as a counselor following that path has strayed far from empathy.

Don't Do Most of the Talking

As your expressions of empathy will be more precise and concise than your clients' communication to you, you should do only a small percentage of the talking when expressing empathy. Further, when you focus your responses on emotions, there is less to say in response than when listening therapeutically to client content. Emotions need not be explained, justified, or proved, and once understood, take few words to express.

Don't Make "Me Too" or "Must Feel" Statements

Sometimes counselors, including us, are tempted to "me too" our clients. By this, we mean that we mistakenly attempt to tell them what we have felt in similar situations or think we would feel in their situation. A sneaky way of making this error is to phrase an expression of empathy as, "That must really hurt." This places an assumption on what the client feels that may or may not be true and is suggestive, thus taking the client away from expressing what she actually does feel. The assumption in this error seems usually to be based on what the counselor thinks she would feel in her client's situation or what the counselor thinks anyone would feel in her client's situation. The assumption may be that we humans would all feel the same things in the same situation. This conclusion is highly unlikely, and this line of thought strays far from empathy.

It is possible, and sometimes helpful as a last resort when you are very unsure what your client feels, to use a very tentative phrasing like, "I imagine that you feel hurt." In this phrasing, the focus is clearly kept on your striving to understand and express what you feel with your client. You may be making an assumption, but that assumption is only about the unique person that is your client and your striving to understand his unique reaction to his situation.

NUANCES OF EXPRESSING EMPATHY THAT DON'T QUITE FIT OUR DO'S AND DON'T LIST

Remaining Animated, Natural, and Spontaneous in Expressing Empathy

We encourage you to allow yourself to be animated in your counseling sessions. What we mean is that we wouldn't want you to think so hard and try so hard to do the right thing and respond the right way that striving to have and express empathy actually gets in the way of your having and expressing empathy. You will need to make a great many judgments about how to respond to your clients that will be unique to given situations, persons, and moments.

Variation of Tone in Expressing Empathy

Generally, you want the level of emotion in your tone to match the level of emotion in your client's tone. Usually, if you are striving to have and express empathy, this will happen without your actually paying any attention to your tone and whether or not it matches your client's.

However, there are always exceptions. For example, you may have a client who repeatedly begins to deny and minimize his emotions when he hears the strength of it in your tone. When this is your hunch, you may try expressing a little less intensity of emotion in your tone than your client actually does, since you already know your client does not accept the stronger tone and in order to allow your client to feel less threatened by what he hears back from you and accept more of his emotion.

On the other hand, you may have a client who seems to assume that she should not express strong emotions, even in her sessions with you. When this is the case, perhaps the best way you can let your client know that it is okay to express strong emotions with you is for you to respond to her restrained expression with a tone that is slightly stronger and more intense than her own.

For us and the many counselors we know who express empathy and facilitate emotional expression well, such decisions have become instantaneous, automatic. At this time in our work, such decisions are more felt than thought. Our point is for you to use our Do's and Don'ts list as it may be helpful to you but not to let it make you rigid. Remember, the most powerful thing you can offer your clients is an empathic relationship with *you*. While we hope following our model helps you become more empathic, we also want you to continue to be you, to remain focused on empathy and trust your own intuition and hunches.

Responding to Implied Emotions

Just as in responding to implied content with therapeutic listening, you will often respond to implied emotions with expression of empathy. Most people, in or out of counseling, do not name their emotions when expressing feelings. An example of responding to implied emotions follows.

A teenage client had a behavior pattern of fighting with other boys, especially when drinking. He related the story of a fight that he had been in the weekend before to his counselor. In that fight, the other boy was hurt very badly. The police had been called, but the client left before they came. As he told of the fight, he explained to his counselor that the other boy must feel terrible, that the way his face had looked, he just could not go back to school and be seen. A reflection of content to his communication would be, "You know how bad this must be for him. You know that he just couldn't be seen with his face beat up." This would be a good and powerful reflection. The client might then respond with more direct expression of the emotion he feels. An alternative response, expressing empathy for the emotion that he already strongly implied could be, "You seem to feel so sorry for him. You seem to feel guilty for your part in his humiliation."

A less clearly implied emotion from this same example is that as this client told of the fight, he sometimes spoke loudly, quickly, and with bright eyes. The undercurrent emotion his counselor felt with him during this part of his story was elation. Empathy could be expressed to this implication by saying, "As you tell me of the fight, you get excited. You seem to feel elated."

Either way of responding to implied emotions—with content reflection, as in the first part of this example, or with stated empathy—could be effective. The most helpful thing was for his counselor to respond fully and honestly to his emotion, whether it be first through tone in a content reflection, giving him the opportunity to express emotion more fully and directly, or through a more immediate expression of empathy. Responding to his emotion helped him realize his emotional experience. It may have been this experience that was most powerful in helping him change. It may have been very helpful for him to realize his guilt and remorse fully in order to avoid repeating such fights. It may also have been helpful for him to realize the elation he had when telling of the fight in order to know fully and honestly what drew him to such fights. Most likely it was not just that he saw himself as justified in fighting (the other boy did hit first, etc), but that it was also a thrill to fight, and a sense of power and mastery to beat up another boy.

Responding to Unpleasant Emotions

It is important to also see from the previous example that in responding to client emotions, counselors need to be ready to respond to unpleasant emotions. Just as these emotions are often hurtful for clients to realize, they can also be hurtful for counselors to respond to. Counselors sometimes avoid them simply because they hurt and humans are naturally repulsed by pain. It may also be that counselors sometimes fear losing themselves in the strong currents of clients' unpleasant emotions. For us and the other counselors who knew of the boy in the previous case example, the emotions around elation in telling of the fight ran counter to our "shoulds," our important beliefs about life, feelings, and behaviors toward others (Our use of the word *shoulds* or *should* as a noun in this or other sections comes from writings by Albert Ellis. To understand

Ellis's meaning of this term, see works such as Ellis, 2005; Ellis and Dryden, 1997; or Ellis and Harper, 1997). In our worldview, it is wrong, unacceptable to derive joy and pleasure from hurting others. Sometimes experiencing the dark side of clients' emotions scares us as it reminds us of our own dark sides, the parts of our selves that are difficult for us to accept (Our use of the term *dark side* owes a debt to the Jungian concept of the *personal shadow*. To better understand this concept, see works by Jung or about analytical psychotherapy, such as Jung, 1935/1956; or Douglas, 2005).

We encourage you to work to make this boundary unnecessary in your work. Steady yourself to experience hurtful, repulsive emotions with your clients in order to be the best tool possible for your clients' self-actualization. Practice, contemplate, and do your own personal growth work so that you are able and confident to experience your clients' strong emotional currents without losing yourself. Work to be aware of your shoulds, so that you can temporarily suspend them in order to experience with your clients. Take on the long-term task of knowing your own dark sides to help you be able to temporarily set aside your convictions of right and wrong in order to experience fully with clients in crucial moments. Remember that *empathy does not equal agreement*, but requires coming as close as possible to shared experience while remaining separate. We offer additional guidance to you for this work in the Activities for Further Study section at the end of this chapter.

Discerning When to Respond More to Emotions
and When to Respond More to Content

There is no sure prescription of how often you should respond with therapeutic listening to the content of your clients' communication and how often you should express empathy to the emotions of their communication. As a rule of thumb, you should respond to emotion with empathy as much as possible. But how much is possible will vary with each moment and each client. The best way to discern how much you may respond to a client's emotions in a given moment is to watch for her response when you do. If it was right to respond more heavily to emotion, your client will likely respond with intensified emotion. If it would have been better to respond more to content, your client may respond to your expression of empathy with something like this: "[After a mildly confused pause] Well yeah, but what was happening was . . ." Such a response can indicate to you that your client sees getting the story out as the most important thing she is communicating to you in that moment. But as with most basic skills, this is not a decision we knowingly make in our work anymore. The decision has become more felt than thought, and so is virtually instantaneous. We expect such decisions for you will also become automatic over time in practice and evaluation. You may use the activities of this and other chapters, the feedback you get from the practices at the beginning and end of the chapter, and the feedback you get in supervision to help you grow this decision process into a natural, felt way of interacting.

COMMON DIFFICULTIES, PITFALLS, AND DEAD ENDS

We would like to share with you the problems that we know to be common from experiences of our own and those of our students and friends.

Thinking of the Word, Rather Than Feeling with Your Client

In our section, "Various Ways to Express Empathy," we explained that naming clients' feelings sometimes stops their communication. When counselors work hard to think of the name of a feeling, it takes counselors into a cognitive realm and away from empathy. When counselors name the feelings that they have thought of, it comes across sounding like a conclusion or the final answer on a quiz show. So instead of fostering clients' further experience and expression, it stops it. Some beginning counselors think too much of expressing empathy as naming the feeling. It is this way of thinking that produces this error. Remember that language is almost always inexact for what we humans really mean to say. If a word happens to come to you that is very close to expressing what you feel with your client, it can be helpful to say it. If not, remember that you have a variety of means to express what you feel with your clients. We explored some in our section, "Various Ways to Express Empathy," but there may be many more. Also remember that naming the feeling can have the effect of clarifying the feeling, which can be useful, but the more important task is creating and enhancing a connection between two people, for the client to know that he is understood.

Trying Too Hard to Get It Just Right

A closely related difficulty is worrying about being wrong or trying too hard to get the right word for what you feel with your client. We remind you that as you strive for empathy, errors are not harmful and are sometimes accidentally helpful. Beyond the usual limitations of language, especially for feelings, which are abstract, there rarely is a right word because feelings seldom exist in isolation. Usually feelings are mixed like flavors in a dish that is not finished cooking. Feelings are ever changing, in constant motion. It is right to try to understand exactly what your client feels, to try to experience right with her. You may accomplish this at times. It is right to try to find just the right way to express what you feel with her, but remember that there will seldom be opportunities to find just the right words to express it.

A metaphor for thinking too much might help. If we learned to ride a bike as an adult (we were more talented at thinking less and simply being or doing as children), we may need to think of where to sit, how to hold on, how to pedal without having our feet slip off the pedals, how to steer, how to break, and so on. At first it may feel quite unnatural. But soon it would be time to forget all the skill pieces and just ride, letting the forward momentum of our base skills pull us along. Yet, as soon as we'd think, "Hey, I'm really getting it. Look at me," our break in concentration of purely being and doing the complex set of behaviors of riding, we just might wobble, lose our balance, and fall. However, if we continue to think too much of each complexity in base skills, we may keep ourselves both from the joy of riding and from the other layers of skills available to us, like jumping ramps, riding on trails, riding through traffic and through different terrain, landscapes, and weather. So like riding a bike, you may practice your base skills until they come naturally and you do them almost without thinking. Also, you may often find your way of being that feels right for you in counseling, that feeling when you know you are doing your base skills well even though you are not having to think about them much; that way of being where you know you are concentrating on what you are doing, but you are much more doing it than consciously thinking about it.

A Limited Vocabulary for Feelings

Related to this, another difficulty that beginning counselors often face in expressing empathy is a lack of vocabulary for feelings. Occasionally our students study lists of feeling words to increase their vocabulary of words for feelings. Some find this helpful. We imagine that if we had tried to develop our vocabularies for feeling words in this way, our focus in sessions might have veered to the list and away from our client. For us, extensive vocabularies of words for feelings have come from being open to and focused on feelings, from desiring to express feelings, and finally from gaining more and more experience in expressing empathy as counselors and expressing our feelings to our counselors, friends, and family, while remembering that words are only one medium of expression.

The Problem with Claiming Understanding
or Shared Experience

Again, just as with therapeutic listening to communication content, it is almost never helpful to say "I understand" or "I feel what you feel" to your clients. When you think you understand something or believe you are experiencing with them, then say what you understand or experience. If you are right, your client will then know you understood or experienced with them. If you are wrong, your client can know you are trying. In life outside of counseling, acquaintances, friends, and family often tell each other that they understand or know what each other feels. In many cases, the person who hears this does not believe it. Such statements from errant counselors often leave us thinking of young teenagers whose parents tell them, "I know just what you're going through." At least when we were teenagers, we rarely believed such statements, even from very caring adults. It seems that what persons often mean when making such claims of understanding is really something like, "I'm pretty sure I know what I felt in situations that seemed very similar to me." Now this might actually be a useful sharing of experience among friends, acquaintances, or family, but it is not the empathy that you can much more powerfully express in counseling.

Personal Confidence and Faith
in the Counseling Process

Beginning counselors are sometimes tempted to stray from expressing empathy due to a belief that goes something like, "Surely my expression of empathy cannot be enough." This is sometimes a lack of faith in expressing empathy and related counseling skills. When that is the problem, experience in giving these skills a solid try and assessing progress will go far toward curing that doubt.

In a sense, empathy is not enough, as it is a core of what counselors do but is not all that counselors do. However, more often we find that doubting the power of empathy in counseling is more about ourselves as counselors than about empathy. It seems to tap more into a core of self-doubt that says, "Surely *I* cannot be enough. This skill is too dependent on me. Surely a technique invented by someone else (somehow smarter or better?) would be more useful than me and my empathy."

Particularly because we know how difficult such self-doubts can be, we'd like to offer some helpful thoughts and suggestions. First, you are not alone. We still sometimes face such doubts, even though we have quite a lot of evidence of the effectiveness of our counseling. Also, the more success we've had as counselors, while trusting the power of empathy, related skills, and ourselves, the less self-doubts we've had. So we expect a similar progression for you. We've also found it helpful not to let ourselves get isolated. We've known ourselves and our friends to work too much so that the work becomes consuming. We find it helpful to back off from that myopic consumption to meet regularly with like-minded counselors, to share our experiences, our successes, and our self-doubts. Further, this book is no doubt only a part of your study. For us, an ongoing scholarship with the core conditions of counseling is continually uplifting. Finally, we have found it very helpful to seek counselors who honor empathy and the core conditions for ourselves. As we are confident in the self-actualizing process, we know that we have continuing work to do in our ongoing self-discovery and acceptance.

A Lack of Unconditional Positive Regard

Another difficulty in expressing empathy can be a lack of unconditional positive regard. A counselor may be so overwhelmed with critical reactions to a client who has done or said something that runs against one of the counselor's moral constructs, that the counselor may have difficulty finding and expressing empathy (If not an action or word, the client may have an abrasive, off-putting way about her). In a seemingly opposite scenario, a counselor may feel so fond of a client that his positive regard becomes possessive, accepting the client only if and when she makes expected, desired progress. This too would inhibit experiencing and expressing empathy. Our next chapter, "Striving for and Communicating Unconditional Positive Regard," will provide you with ways that you may work through difficulties in experiencing unconditional positive regard that limit your expression of empathy. We hope you will see that empathy and unconditional positive regard work in tandem to support each other and all counselor actions.

ACTIVITIES AND RESOURCES
FOR FURTHER STUDY

- Our Do's and Don'ts list is our best attempt to boil down the behaviors of expressing empathy to just the bare skeleton of what happens. We add narration to explain the behaviors. Still, the behaviors will come naturally to you sooner if you create understandings that are most meaningful to you. So, through class or small-group discussion and private contemplation, first check that you understand the Do's and Don'ts list items, then narrate or explain them in your own words. Once you have your own explanations, you may choose to rename some of the Do's and Don'ts based on how you phrase what they mean to you. You will then have your personalized Do's and Don'ts list for expressing empathy that may become most automatic in directing your behavior in counseling sessions. After working out your personalized list, you may then return to class or small-group discussion to explain the adjustments you have made, hear from others, and benefit from the shared perspectives.

- Review the instructions from the focus activity. Then, practice expressing empathy under similar circumstances. However, now that you have read and contemplated this chapter, practice expressing empathy following the guidance given in this chapter. When giving and receiving feedback from partners and video, check for adherence to the items of the Do's and Don'ts list. Also check for the various ways of expressing empathy that seemed to work for the listener in each practice round. Repeat practice rounds until each partner is reasonably satisfied with her or his adherence to the Do's and Don'ts, percentage of accuracy, tone, and way of being in expressing empathy.

- Discuss possible hidden agendas (or overt agendas) that might tempt or motivate you away from expressing empathy as counselor and ways that these agendas may affect your work. Don't be shy about admitting such temptations. They would not be so much wrong as normal. Being with persons in pain is not easy, and sometimes much of our background prompts us away from expressing empathy. Here are a couple of examples to get you started: seeing a client suffer with painful emotions and inadvertently holding yourself back from expressing empathy in the hope that she will not have to feel her pain so fully; or feeling compelled to give a solution that you think is simple for a client who is in pain (especially a client who you see as experiencing a difficulty you have recently faced) and believing that this advice might help your client avoid pain, when expressing empathy might actually intensify her pain.

- Hold a discussion of the various difficulties that you know or imagine will affect you in expressing empathy.

- Give special consideration to the difficulty counselors face in experiencing and expressing empathy with unpleasant emotions. Speculate on the difficulties you will face in that part of your work. Plan ways to overcome your limitations in that work. Notice your reactions to unpleasant emotions in life, and real and practice sessions. Contemplate the meanings of what you notice in your reactions for yourself and in your work. Design practice sessions to challenge the weak points that you identify. Use counseling opportunities and contemplation of life experiences to discern your shoulds, so that you can knowingly set them aside when need be. Use counseling and other opportunities for deeply honest, authentic human interactions in which you can risk expressing and coming to know yourself, in order to come to know and accept your dark sides, so that you need not fear them and need not limit your experiencing with clients.

- Journal your thoughts, feelings, and understandings of self and the subject of expressing empathy, especially as they result from these practices and discussions. Give yourself time to contemplate the main ideas of this chapter and your reactions.

- Review the Primary Skills Objectives and your satisfaction of your mastery of each at this point in your development. Use your review to guide you to the rereading, additional reading, and additional practice you seem to need.

5

Striving for and Communicating Unconditional Positive Regard

All sentient beings should be looked on as equal. On that basis,
you can gradually develop genuine compassion for all of them.

DALAI LAMA

Nothing we do, however virtuous, can be accomplished
alone; therefore, we are saved by love.

REINHOLD NIEBUHR

PRIMARY SKILL OBJECTIVES

- Understand and develop an explanation that can be understood by your classmates as well as by nonprofessionals of what unconditional positive regard is and why it is important, powerful.
- Understand and be able to explain how unconditional positive regard is and is not expressed, and how it is and is not perceived, and why.
- Understand and be able to explain the applications of unconditional positive regard in life in general, specifically in your life, as well as in your present and/or future counseling sessions.
- Understand the ways that major authors on unconditional positive regard cited in this chapter have explained unconditional positive regard.

- Understand and be able to explain the inhibiting factors for unconditional positive regard that are discussed in this chapter, and be able to explain the factors that may especially inhibit your having and expressing unconditional positive regard.
- Be able to explain the likely relationship between counselor burnout, difficulty expressing unconditional positive regard, and the potential effects of this counselor difficulty on clients.
- Outline your plan for avoiding the burnout that may inhibit your unconditional positive regard and negatively affect your clients.
- Describe your strengths and weaknesses in providing unconditional positive regard.
- Discern how you may maximize your strengths and improve your weaknesses in providing unconditional positive regard.

FOCUS ACTIVITY 1

There is a quote from Carl Rogers that we find quite true and profound, but at first it seems to not make sense. It is an important paradox to contemplate and resolve for understanding therapeutic relationships and unconditional positive regard. The quote, in two parts, is "When I accept myself as I am, then I change," and "We cannot move away from what we are, until we thoroughly accept what we are. Then change seems to come about almost unnoticed." The paradox to us is that it seems logical that if people accepted themselves, they would stagnate, lose motivation to change. Yet, we do not find this to be true in others and in ourselves. Contemplate, speculate, journal, and discuss how Rogers' observations could be true, despite seeming not to make sense. Work at it until you have a tentative explanation that seems meaningful to you. Apply your speculative answers to yourself and to clients whom you know or imagine.

FOCUS ACTIVITY 2

Journal and then discuss your thoughts on the following sets of questions with a group of your peers. Answer as many of the questions as possible. Don't allow yourself to quit easily on these difficult questions. A partial answer will be better than none. You can build your answers in discussion. Take time to contemplate, rather than answering quickly. Ponder and look for themes in your answers and those of your discussion group.

Additionally, in case they begin to, try not to let your answers prompt self-criticism. There are no wrong answers, and consistent positive regard may be more rare than some might think. Whatever your answers are is simply a part of who and how you are at this time. Our sets of questions to ponder, journal, and discuss are not meant as a means of right/wrong, good/bad evaluation. Also, for this focus activity we refer to consistent or frequent positive regard, instead of using the term *unconditional positive regard,* which we will define in the chapter.

1. Describe the persons or things that you almost always have positive regard for:. What and/or who are they? How did they come to be in your life? How has it happened that you hold them in such frequent positive regard? What sets them apart from the persons or things that you have less consistent positive regard for or that you have negative regard for? How do you feel, think, and act when you hold them in positive regard? When you do not? How do you feel, think, and act toward persons for whom you have much less frequent positive regard?

2. Describe your relationships with persons, if any, who almost always hold you in positive regard: Who are they? How did they come to be and remain in your life? How has it happened that they hold you in such frequent positive regard? What sets you apart from the persons or things that they seem to hold in less consistent positive regard or hold in negative regard? How do you feel, think, and act in response to receiving nearly consistent positive regard? How do you feel, think, and act differently when this positive regard is not there, or when negative regard is?

3. Now think of positive regard in relation to yourself: How consistently do you hold yourself in positive regard? On what does the positive regard in which you hold yourself seem to be based (e.g., number of and closeness with loved ones; just being; a spiritual conceptualization of yourself as part of a higher, more perfect being; level of accomplishments; some combination of these and other factors)? What does the negative regard you sometimes have for yourself seem to be based on? How do you feel, think, and act in response to the times you hold yourself in positive regard? In negative regard? Does your reaction seem to vary when the positive or negative regard seems to be based on different motivations? How so?

From your considerations: (1) What do you conclude? (2) Is there a connection between giving and receiving positive regard? If so, what? (3) Is there a connection between positive regard to or from others and with yourself? If so, what? (4) Is it difficult to give and receive consistent or frequent positive regard? Why do you think this is or isn't? (5) What do you think your difficulties will be in holding clients in positive regard and how might you improve?

INTRODUCTION

Paths toward Holding Others in
Reasonably Consistent Positive Regard

Having taken ourselves through the focus activity, we are reminded that consistent positive regard can be quite difficult. We find that outside of counseling it can be difficult for us to give and to receive positive regard consistently. We have found it easier to hold clients in positive regard than some persons outside of counseling, and sometimes ourselves. Part of what has led us to most frequently hold our clients in positive regard is connected to our having devoted a large portion of our work to counseling children, especially early in our careers. We tend to see children as innocent and closer to the natural world, perhaps closer to the higher power that we believe we are all a part of, and which, in the unity of life, is more perfect than any individual. We also clearly see that children are constantly, instinctually learning; self-improving; and, except for massive interruption, rapidly self-actualizing. In counseling persons of various ages, we have continually broadened the breadth of persons that we see encompassed by the innocence of a child. Our definition of whom we hold in this same positive regard as children has continually widened as we worked.

To see that our clients are each instinctually striving to learn, self-improve, and self-actualize has helped us to hold consistent (for the most part) positive regard for our clients. This is easily seen in cases where clients come to us or other counselors and say, "I want to use counseling with you to better myself." But we also find this true in the much more common cases where the client was referred by a concerned person (i.e., a teacher, parent, spouse, friend) or, as is often the case in schools, was invited to use counseling by a counselor, even in the cases where the persons were quite reluctant to be in counseling. In all these cases, we have found that if we look for it, we can see that each person is striving to self-actualize. We also find that engaging our clients with the core conditions of counseling tends to help us see their striving to self-actualize and also seems to strengthen this striving. So with counseling, a self-perpetuating cycle begins that the more we meet clients with the core conditions, the more they strive to become their best and the more we can see this striving and the easier it is to hold them in positive regard. Then, the more easily we can hold them in positive regard, the more their natural striving is strengthened, and on and on.

As we have already implied, we sometimes find our way back to positive regard for others and for ourselves, after having strayed, through our spiritual beliefs. We prefer to keep our comments on this few and as open to your spiritual beliefs or thoughts as possible. We do not mean to imply that a spiritual connection with positive regard for clients is necessary or within every counselor's path. Still, other authors have also noted a connection of unconditional positive regard to well-being and growth in the tenets of major religions, "For example, Purton (1996) equates the Buddhist concept of counteracting *lobha,* or greed, with the development of unconditional positive regard, and 'forgiveness' and compassion are at the heart of Christian (and other) belief." (Wilkins, 2000, p. 25).

The Tandem: Empathy
and Unconditional Positive Regard

We think of unconditional positive regard and empathy as a tandem—a tandem as in a two-wheeled carriage pulled by two horses, one before the other, or a tandem bicycle (The American Heritage Dictionary, 1976). They are linked, difficult to separate, and always much stronger together. They are mutually giving partners. It is impossible to say which, if either one, leads the other. In our work, empathy has been the key to making unconditional positive regard possible in difficult moments. For many counselors, refocusing their self in unconditional positive regard helps them renew their strength in empathy.

As with empathy, Rogers was the first to introduce the helping professions to the major importance of unconditional positive regard in a major way (Rogers, 1957). Like empathy, the concept of unconditional positive regard can seem esoteric, and it can be difficult to grasp just how to provide it. We can't promise to tell you just how to provide unconditional positive regard because that process can be unique to each person. However, one of our goals with this chapter is to make unconditional positive regard as clear and tangible as possible. Another is to guide you to find ways to maximize the unconditional positive regard you provide and to be well on your way to adding this essential element to your therapeutic relationships.

WHAT UPR IS AND ISN'T

Beginning Thoughts on What
Unconditional Positive Regard Is

Rogers associated the terms *warmth, acceptance,* and *prizing* with unconditional positive regard (UPR) (Wilkins, 2000). Rogers (1961) explained that UPR is "a warm caring for the client—a caring which demands no personal gratification. It is an atmosphere which simply demonstrates 'I care'; not 'I care for you *if* you have behaved thus and so'" (p. 283). Mearns and Thorne (1988) explained, "The counselor who holds this attitude [UPR] deeply values the humanity of her client and is not deflected in valuing any particular client behaviours" (p. 59).

Of acceptance, Rogers (1980) wrote, "The therapist is willing for the client to be whatever immediate feeling is going on—confusion, resentment, fear, anger, courage, love, or pride" (p.115). So the acceptance is just as true for emotions that the counselor might see as negative or as positive in situations outside of counseling. For example, if a counselor outside of sessions finds others' expressions of anger distasteful, in sessions this counselor will need to work past that distaste in order to accept all of the emotions her clients express. If we believe persons should feel guilt for their actions that hurt others but a client in a session expresses elation rather than guilt over such behavior, in order to be helpful, we must accept and respond with empathy to this expressed elation. We have known clients to begin to reevaluate their reactions and change their hurtful behaviors with counselor acceptance of their feeling of elation. We have known clients to hear their own elation clearly when it is met with warm acceptance and empathy, then to find it distasteful and begin a process of realizing how they have been motivated into hurtful behaviors. However, an important paradox to realize is that when faced with such a situation, you cannot meet this elation with warm acceptance and empathy in order to have your client change. If you did so, this would be your agenda for your client and not true acceptance. Such a reaction would either thwart the change or bring a superficial change at best—a change that is meant just to please you and may last no longer than your presence. Such an agenda-laden response might also prompt guilt without further self-discovery, self-acceptance, and self-empathy, and then enhance a cycle that drives hurtful behaviors in your client.

We like the use of the word *prizing*. It suggests to us that clients are treasures, that it is a joy and pleasure to discover them. We think of the way parents often prize their new baby, "Oh, look—another poop! He's so healthy," "Feel how she clasps. So strong!," "Uh-oh, he's tired. Time to go lay down." Such prizing is often infectious to clients and helps them find the strength to renew and restart or enhance their drive to self-actualize.

We do want to emphasize but not overstate this quality of prizing. Prizing isn't infatuation. But it can be loving and appreciating the uniqueness of each client. It can be enjoying discovering our clients as they discover themselves. It can be appreciating both the big dramatic parts of their discovery, which are really quite rare, and the mundane details and little nuances, which occupy most of the discovery time in counseling.

We also like the reference to UPR as prizing because it is an action verb. UPR has been an active and ongoing process for us within the relationships we have had with the clients we have served. Our clients have often allowed us to see the more abrasive or harder-to-like parts of themselves at that time in their life. We certainly hope that each new client will allow us to see these aspects of himself, as these are likely important parts of himself and aspects that may not easily meet with acceptance in other relationships.

It is important to note that UPR does not mean agreement or approval. We have counseled persons whose behavior we would not have approved of outside of counseling. Oftentimes, through counseling, those clients chose to change those outside behaviors that we would not have approved of. We have counseled persons when we strongly disagreed with some of the things they were saying. Our work in child-centered play therapy (CCPT) (Guerney, 1983, 2001; Landreth, 2002) has been helpful here. In CCPT, play is the child's primary language, not words. From this, we have been reminded that when adults in counseling use words that are upsetting to hear, those words are a means of expressing who they are and how they feel in that moment, and that the words are less important than the person saying the words. An example from Jeff's work:

> I had an adult client who frequently made racist statements and harsh-sounding statements about what he considered liberal ideas or persons. I sometimes wanted to argue with these statements or at least tell him that I disagreed and that I hurt to hear them. He seemed to be beginning to show me his abrasive side, and I guessed that this part of him had been formidable in driving others away. Rather than focusing on the content of those words, I continued to hold his person in positive regard, even when I disagreed with his words. His words were more important in counseling as a manifestation of how hurt and bitter he felt than for the content they conveyed.

We also like the use of the word *acceptance* to help understand the meaning of UPR. Acceptance certainly seems less positive than prizing, but sometimes it is the most we can muster. Acceptance suggests the existential notion that persons, things, and situations simply are what they are. Knowing that they are what they are, that they are the best they can be in that moment, doesn't mean that we might not wish for change or that they will not change. Rather, change is inevitable. For example, if the day is cold and gray and I wish it were sunny with a dry breeze, I will only frustrate myself wishing it were different. My wishing will not change the moment. If I accept the current weather conditions, I have the best opportunity to experience what that day has to offer and to be better for it. That way of being can then bring me closer to prizing the day. If we accept each client in his current condition, we give our clients and ourselves the best opportunity to accept what they have to offer, and they and we are the better for it.

To us, the use of the word *acceptance* also suggests part of our belief that we are all part of a greater whole, a whole that is more perfect than any of us are separately. For example, in our geographic area cold, gray days in winter are required to prepare for spring. Spring must come before summer, and the heat of summer helps us appreciate fall. Our point is that acceptance partly means to us seeing each person as a part of a whole. If that part is hard for us to accept in that moment, we can help ourselves by remembering that they are a necessary part of the whole, as are all things.

Like empathy, we believe that UPR is natural to all humans. So, learning to provide reasonably consistent UPR for your clients will not be a task of learning a new skill but a task of discovering an old way of being that may have gotten squelched by personal inhibitions, social pressure to compete, or other factors. When award-winning kindergarten teacher Vivian Paley uses the words *kindness* and *interest in others,* she describes what sound to us like a universal UPR that is a strong and natural force among children (Wingert, 1999). This inspires and reminds us that UPR is natural to all people.

What UPR Is Not

Wilkins (2000) helped clarify UPR by reversing each of its elements. This way of considering what UPR is not may help you more fully understand what it is. Consider the following carefully.

Conditional positive regard would mean "offering warmth, respect, acceptance, etc. only when the other fulfils some particular expectation, desire or requirement" (Wilkins, 2000, p. 25). We have known a great many persons in and out of counseling whose parents raised them with conditional positive regard as a primary motivator. We have never known it to have a positive effect on the persons' sense of self-worth. For such persons, the result often seems to be a sense of self-worth that is based only on continually compiling accomplishments, manifested in behaviors that we consider maladjusted, such as taking hurtful advantage of others or becoming so discouraged that self-development is greatly thwarted.

In counseling, holding conditional positive regard for a client would mean the counselor has a particular agenda for her (Wilkins, 2000). Thus, the counselor's behavior communicates, "I will accept you *when* you make progress in a particular direction." As with a parent using conditional positive regard as a primary motivator, such counselor behavior would likely only produce temporary, elusive progress at best. Any progress that occurred would then be based on conditional self-acceptance and risk a high level of dependency on the counselor, a crash to a greater low for the client, and inhibited client decision making.

Unconditional negative regard would communicate to the other person, "Whatever you say or do, however you are, I will hate and despise you" (Wilkins, 2000, p. 26). In its extreme forms, it is the root of racism, homophobia, sexism, and the like. While we certainly hope not, we suppose that unconditional negative regard could creep into a counselor's work in its more passive form when a counselor has become burned out on service to a difficult, perhaps abrasive population and is dangerously unaware of his unconditional negative regard, while believing himself to be and perhaps going through the motions of being warm, accepting, and caring.

Unconditional positive disregard would mean that one person totally refuses or neglects a relationship with another. "In an extreme form, it is the complete negation of the existence of one person by another. This can be so powerful that receivers of it come to doubt their right to life" (Wilkins, 2000, p. 26). We have theorized that children with conduct disorder have perceived something like this from their parents at a key stage in development (often the parents were preoccupied with their own difficulties, depression, dissolving marriage, substance abuse, etc). These children then developed a set of self-statements that include such thoughts as, "No one will like or love me. Therefore I am unlovable." In the magical thinking of childhood, this sometimes manifests in a fear that "I may cease to exist. I may have so little impact on my world that I will disappear." Such children then act out in a mistaken attempt to reprove their existence, reestablish a personal power, and drive others away before experiencing a rejection that they believe will be inevitable and intolerable (Cochran & Cochran, 1999). "A distressed, bored or unengaged therapist is in danger of offering unconditional positive disregard" (Wilkins, 2000, p. 26). And as with our theory of conduct-disordered behaviors, unengaged interaction can be interpreted by persons in vulnerable positions as a major and painful rejection.

In summary, UPR is not an agenda (whether known or unknown) to accept only parts of the person, her emotions, or the content of her expressions. UPR is not dislike. It is not a judgment of right, wrong, good, or bad. It is not agreement or approval. It is not disapproval or disengagement. It is not going through the motions of liking or merely acting as if you want to be with the other person. It is a warm, caring, nonpossessive acceptance of the person, her emotions, and the content of her expressions. It is an affection for and a prizing of that person.

A SAMPLE OF THE LITERATURE
SUPPORTING AND CLARIFYING
THE IMPORTANCE AND POWER OF UPR

There is, of course, strong evidence supporting the power of the core conditions of counseling as a set (e.g., Bergin & Lambert, 1978; Krumboltz et al., 1979; Orlinsky & Howard, 1978; Patterson, 1984: Peschken & Johnson, 1997) and still more regarding UPR as a singly important factor. Wilkins (2000) explained that "the communication of UPR is a major curative factor in any approach to therapy" (p. 23), and it is *the* curative factor in the context of congruence and empathy. Orlinsky and Howard (1978) concluded that patients' perception of therapists' manner as affirming the value of their person is significantly associated with good therapeutic outcome. Chiu (1998) concurred, explaining that UPR is the context to his holistic biopsychosocial approach. Jacobs (1988) recorded that psychodynamic counselors would expect "unconditional regard" to encourage positive transference (p. 13). Bozarth (1998) described it as the curative factor for client-centered therapy. From their meta-analysis, Farber and Lane (2002) stated that "positive regard seems to be significantly associated with therapeutic success," especially when judging by clients' perspectives on positive therapeutic outcome (p. 191). They explain that, at a minimum, it "sets the stage" for other positive interventions and, at least in some cases, "may be sufficient by itself to effect positive change" (p. 191).

Regarding the power of UPR in relationships outside of counseling sessions, Peacock (1999) included UPR as a primary quality needed in custodial interactions in therapeutic communities; van Ryn and Heaney (1997) as an essential component in health education practice. Cramer (1994) evidenced that self-esteem is determined by the degree to which one has a close friend seen as providing the core conditions, especially UPR; and Scheffler and Naus (1999) that a perception of UPR from fathers predicts self-esteem and a lack of fear of intimacy in daughters.

WHY UPR IS IMPORTANT, POWERFUL

We hope that the focus activity of this chapter was helpful to you in beginning to discover some of how and why UPR is important and powerful. One way to understand how and why it is powerful is to know some of the ways people respond to experiencing warm, heartfelt positive regard from others. We humans often have a variety of responses to experiencing this positive regard, including coming to see ourselves in a more positive regard; coming to accept ourselves; feeling empowered; feeling encouraged to learn more about ourselves and who, how, and why we are the ways we are; and feeling empowered to risk going into scary places of self-discovery. Jeff would like to offer a hopefully instructive example of his reaction to UPR.

Though it sounds a little odd to me, I sometimes have a twinge of guilt when I realize that others truly accept and hold me in positive regard. This seems to be connected to my thinking that I don't deserve it, mostly because I don't hold others, even those who frequently hold me in warm positive regard, as consistently in the warm positive regard that I think I should.

Even when a client has a reaction like this to UPR, and his counselor meets this reaction with deep, heartfelt empathy, this can be very useful to the client. This can be the opening of the door for that client to reconsider his self-statements, to review who he really is, his standards for himself, and what he wants those standards to be. Such a client reaction can be quite painful but a moment of opportunity to begin great progress through the process of counseling focused in genuine empathy and UPR.

Self-Acceptance = Change

Rogers (1961) wrote of a paradox of UPR, "When I accept myself as I am, then I change," and "We cannot move away from what we are, until we thoroughly accept what we are. Then change seems to come about almost unnoticed." (p. 17). We have pondered the meaning of this paradox alone, with our friends, and with our students. From our contemplations, we have a few thoughts we'd like to offer. Accepting one's self requires knowing one's self. If we do not know ourselves, we cannot know who or what it is we are or would like to change. Yet, if we humans do not accept ourselves, we will not allow ourselves to know who we are. Each of us humans will keep the parts that we might imagine are flawed, embarrassing, or shameful hidden from ourselves in the darker recesses of our inner selves and not let them out into the light of conscious examination. But once we do let all of our true selves into the light of conscious examination, we will then either reduce the self-expectations that we find unreasonable and hurtful or change the actions and ways of being that we truly do find unacceptable.

At this stage of counseling, the force of change is incredibly strong and the counselors' work is so easy it's almost obsolete. If a client asked us to guide her behavior through life in a stage of counseling before such self-acceptance, we would not usually give this guidance (exceptions would be situations of imminent danger or greatly destructive behavior patterns that might prevent safe passage into deeper stages, which we discuss in Chapter 10), even if we possibly could. Doing so would likely prevent that person from further self-discovery and self-acceptance, and may help her defer impetus to true, deep, and lasting change.

On the other hand, once reaching a stage of true self-acceptance, our clients normally find the personal skills they need and have come to want readily, without our moving into a teaching mode. If in that stage our clients did ask for help with learning skills for the changes they want, we could more freely offer to teach and do so without undue, change-inhibiting influence.

UPR = Full Expression of Emotions

You may remember a story from the last chapter, under "Responding to Implied Emotions," of a boy who told his counselor of a fight he had been in. A part of his full and honest telling included his expression of elation regarding the fight. Had his counselor not provided him with a relationship based in a large part on her UPR, he would have censored that part of his story, if he told the story or continued in counseling at all. Fortunately, she did hold UPR for him. His full realization of his elation seemed a significant part of his discovering who he was at that time, what parts of his self he was dissatisfied with, and who and how he really wanted to be.

As We Accept Our Clients,
Our Clients Come to Accept Themselves

Modeling is a strong force within counseling based on the core conditions. It takes great effort from a client not to accept herself in the face of a counselor who knows her well through deep empathy and provides her with nearly consistent UPR. A related example from Jeff's education illustrates this.

> Some years ago I was counseling and learning under the influence of a group of teachers and supervisors who were most interested and skilled in cognitive–behavioral approaches to counseling. My primary supervisor and I were glad to see my clients making progress in changing the self-statements that drove their misbehaviors and the emotions that they considered negative. These clients were studying at a prestigious university, and their self-statements tended to be highly critical. She asked me, "So, what do you think it is about your work that is most helpful to your clients in changing these critical self-statements?" At first I wasn't sure, so I pondered and observed my work with my clients further in search of an answer that satisfied me enough to present to this supervisor that I greatly admired. I was expecting to credit something from cognitive–behavioral technique. But what I concluded seemed too simple to be profound. I discovered that they seemed most helped in changing their self-statements by the fact that I didn't try to force them to change these self-statements, but I also simply did not agree with their critical self-statements. I created a safe and empathic environment for each client to let me know who she really was. And as I knew the person that each client really was, I accepted and prized that person. I decided it was my UPR that most helped my clients change their critical self-statements.

A Safe Environment

Both empathy and UPR are essential ingredients for producing a safe environment for self-exploration and self-revelation. If we were clients to counselors that we feared were judging or diagnosing, as diagnosis can often be taken for or used for judging a person as flawed, we would not feel safe and would be guarded from sharing parts of ourselves that we might see as shameful or embarrassing.

But UPR and this safe environment are more than just not making major errors in judging clients. When people have opportunities to bask in the glow of those who care for us and hold us in very frequent positive regard, we *want* to tell them more and more of ourselves. We want to have those caregivers really *know* us.

Evaluation by Others Can Be a Poor Guide for One's Self

Experience is the best teacher. Yet, it often takes UPR to build that place of safety to stop denying the experience of our emotions, to risk experiencing emotions fully, to discover who we really are, what we really want, and how we really want to be. Then with UPR, deep empathy enhances, clarifies, and brings clients' emotional experience into full consciousness. We have known clients in this context to make decisions that they felt great confidence in, and looking back, we agree that these decisions were mature. We have counseled persons who were acting out with misbehavior, experiencing great somatic pain, or felt greatly depressed, discouraged, and unmotivated for their life's tasks but did not know why. In the atmosphere of UPR and deep empathy, they discovered that they had internalized life paths that they had believed others wanted them to take and that they were deeply dissatisfied with these paths. Yet, until motivated to counseling by their troubling symptoms, and until finding themselves through UPR and deep empathy, they had not realized their level of dissatisfaction with their current paths.

Having now made comments upon the importance of UPR, we want to insert a reminder that valuable work in counseling is rarely simple. We don't want to imply that it is simple by telling only the beginning and end, as in the examples just described. Such self-discoveries of our clients through UPR and deep empathy sometimes occur quickly but are almost always the product of hard work and perseverance.

Rewards for the Counselor and the Client

Rogers (1980) wrote, "I have found it to be highly rewarding when I can accept another person" (p. 20). He explained that each of us are islands unto ourselves, in a very real sense. We take this to mean that, while it is right and helpful to strive for it, we humans cannot reach true and complete understanding of each other. The best persons can do is understand what we see of each other through our worldview, formed of our unique set of experiences and meanings made of those experiences. This understanding, and especially our communication of it, must then pass through some medium (e.g., language, play, art), which brings limitations to the connection. But when we feel safe enough to and are permitted through UPR and shared emotion to be and express who we really are, we begin to experience moments of connection. When we counselors accept the feelings, attitudes, and beliefs that are vital to who our clients are in that moment, we help them build their connections, initially to their counselor but also to all life beyond their self. Rogers explained that this work, this connecting, is highly rewarding to him. It has also been highly rewarding for us.

We have often thought that our clients and our students bring out the best in us. We find that when they strive to be who they are with us, and through UPR come to be the best of their potential, this brings us a warm joy and a contentment with ourselves and with them. So, from that feeling grows a desire and capacity to provide more and deeper warm acceptance and prizing of these persons and others. This process has greatly improved who and how we are. We are thankful for it and value it greatly.

HOW UPR IS AND IS NOT COMMUNICATED

It would be difficult to tell a client that you feel unconditional positive regard for her, or that you like, love, or warmly accept her. In life, such things are sometimes said but either are not fully meant, or are misinterpreted or disregarded. Sometimes a part of what brought persons to counseling is that they have difficulty hearing and believing such statements. In counseling, the need to explain such statements might steal any power that such statements could have. For example, if one of us had a client that we struggled to like, whose statements or even expressed emotions were distasteful, but we were successfully working through the distaste to achieve a level of warm acceptance for this client, to state the acceptance would also seem to require explaining what it is based on, which could be or sound conditional, or at least impersonal.

Additionally, if it became your habit to state your UPR in moments when you realize it, your clients could come to assume that if you don't state your positive regard for them, that you either don't feel it or you don't realize it. Further, if you would state your moments of positive regard, would you be willing to also state your moments of negative regard for your clients? We would not normally be willing to do this and know that doing so would take our focus far from empathy, therapeutic listening, and helping our clients self-discover and self-evaluate.

Fortunately, when you feel UPR for your clients, it will usually be clear to them without your stating it. In the less common situations in which your clients do not perceive the UPR that you feel for them, their misperception of how you feel toward them, how they think you see them, will come to light through therapeutic listening and empathy. Then, their perception of how you see them will likely be important for them to learn about. This same sort of misperception may be problematic for them in life.

An analogy from romantic relationships may be helpful. When people are attracted to others, the person they are attracted to usually know this at some level, especially if the two persons have frequent contact. To express this attraction, the people who are attracted need only be at a level where they can be themselves and accept how they feels toward others, versus being greatly inhibited. The people receiving the attraction could probably figure it out by interpreting their admirer's actions, but this is a highly fallible process. Instead, if we humans are open to it, we know without thinking when we are liked. If we are inhibited from knowing this, that would be something worth working on in counseling, and this work would occur through experiencing UPR, empathy, and a therapeutic listener who can respond to the little clues that the client is having difficulty perceiving acceptance or liking from others. One such little clue to a therapeutic listener might be a statement like, "Well, you'd be bored if I tried to tell you about that." To such a statement, we would express awareness and acceptance of her inhibition by reflecting with genuine warmth and mild surprise, "Oh, so you thought to tell me about that but decided to discard it on the mere assumption that I would be bored." Then depending on whether this was a pattern and on how she reacted to the reflection, we might add, "I'd like you to tell me the things that seem important to you, and I'll be interested, as I am interested in you." Another little clue might be a client who seems so concerned with saying the right thing that he has difficulty getting started saying anything at all. To him, we might tenderly reflect, "You seem to worry over finding something good or profound to say." Perhaps, then, a useful discussion of the topics he is discarding will ensue. Accepting, prizing, and attending to clients' experiences and communications (or lack of verbal communication) communicates your UPR and, when needed, opens the door to clients discussing with you their perception of your perception of them (In Chapter 8, "Initial and Ongoing Structuring of Therapeutic Relationships," we discuss helping clients understand how they can most effectively decide what to discuss in counseling, when necessary).

A clear way to know that you are expressing UPR for your clients is to review your work to check that you attend to all client expressions equally, based on the level at which they are expressed to you. The only factor influencing which expressions to respond more to should be your perception of which expressions seem most important to your clients.

One manifestation of UPR is contained in how you accept corrections from clients. As discussed in Chapter 4, under the section "Be Prepared for and Accept Corrections," clients will sometimes correct your reflections in therapeutic listening or expression of empathy. In such a moment, it is not important to convince them that they misunderstood you. To hear their correction and strive to understand that new communication demonstrates both your primary intention to understand as well as your acceptance.

In case it is still difficult for you to discern whether or not you are expressing your UPR for your clients, we would like to offer you some thoughts regarding how you may know if your clients are perceiving UPR from you. UPR tends to bring about a sense of calm, a feeling of being at peace with oneself. This shift is most obvious with children and less obvious with adolescents and adults who may have more greatly solidified and inhibiting self-statements that say things like, "Others will not like me," or "I am unlikable, unacceptable," and who tend to be more guarded against deep, close connections with others. So with children and sometimes with adolescents and adults, you will see their breathing become more abdominal, see their shoulders relax, see tensions fall off their body. When you see this, what you are seeing is your client experiencing the physiological sensation that all is safe and that they are okay with you. This would not mean that your client does not have many more strong emotions to express. When experiencing this safety, child clients go most efficiently to the places they need to go in that moment or the ways they need to be. They may become wildly silly or wildly aggressive in their play. This is your client letting his guard down, letting defenses drop. This is your child client letting self-imposed and societal restrictions drop because at a level close to his core, he feels freedom from judgment and warm acceptance from you.

WHAT GETS IN THE WAY OF HAVING AND COMMUNICATING UPR

Considering common problems in providing UPR is not only helpful to prevent those problems but also helpful in furthering your understanding of what UPR is, and how and why it is essential. Please consider the following discussions of inhibiting factors of UPR.

Having an Agenda for Your Client

Any agenda you may have for a client amounts to placing a condition on your positive regard for your client. Wanting too strongly for clients to change, to improve, is an agenda and inhibits change and improvement. Believing or hoping that just maybe the process of counseling and your client's self-actualization through counseling can be hurried lengthens these processes. Ironically, the less you try to hurry the process, the more efficient you become at helping. Yet, not hurrying for the sake of efficiency is also an agenda. Strive for UPR, deep empathy, and therapeutic listening because they work and because it is the right, most respectful thing you can do. Efficiency will follow.

We do not mean to say that your client may not have an agenda or goal for his own development. This is different from your having an agenda for your client. We discuss helping clients understand how the process of counseling may or may not help them reach their goals in Chapter 8.

We believe that having no agenda for clients, even for them to improve as we or you may see it, may be impossible. Yet, we know that to be most effective, we need to strive to have no agenda and to be totally nonpossessive in our way of accepting clients. The closer we come to this way of accepting, the more efficient we can be.

Counselors Believing They Know Better Than Their Clients

Some counselors approach their clients as if they (the counselors) can see and fully comprehend the complexities of each client's inner and outer worlds. They approach them as if they have analyzed their clients' internal thoughts, beliefs, fears, and motivations, as if they have assessed the realities of clients' outer worlds, which would include the inner worlds of the people that their clients' lives contact. From this vantage point, these counselors seem to assume that they are able to know which behaviors their clients should employ in order to have the greatest success and happiness. This approach also assumes the counselor knows what success and happiness would be for each client and that the counselor's conception of success and happiness is the same as that which would satisfy each client. Such counselor beliefs then lead to conditional acceptance at best, communicating, "I will accept you when you make the changes I prescribe."

Please pardon our slight hyperbole and mild sarcasm in working to make this error clear in the previous paragraph. Probably counselors who take a well-meaning, heavy skill-teaching approach, without a therapeutic relationship based on the core conditions and especially without full acceptance, do not knowingly intend the assumptions we characterized. More likely, in hopes of being quickly helpful, they come to see life and people as much simpler than we do. To us, people and their problems rarely boil down to a need for a discreet set of new skills. In our interactions with counselors who work this way, we find that given time in this way of working, they become discouraged, frustrated, and burnt-out, and come to fill more and more of their work time with noncounseling tasks.

Rogers (1961) explained that he found himself wishing to rush in to "fix things" (p. 21). However, the more he listened to and learned of himself, and the more he tried to extend that same listening to others, the more respect he came to have for "the complex process of life" (p. 21). So, through listening, he became less and less inclined to hurry in to fix things, to mold people, to manipulate and push others to be the way he might have liked them to be.

Burnout

What Burnout Is When counselors experience burnout, they often feel discouraged, frustrated, and bored with their work, and they have great difficulty connecting with clients. We believe burnout usually comes from caring greatly but finding one's self or one's work ineffective. This ineffectiveness could come from poor training or it could come from the ways of being just described in the previous section that interfere with UPR. This sense of ineffectiveness could also come from serving very difficult client populations. In such cases, the counselor's work may be excellent, but still client progress is small. There can be forces in life stronger than therapeutic relationships. Burnout may also be caused or accelerated by a lack of self-care (e.g., neglecting to rest, refusing support, basing self-esteem on amounts of helpfulness to others).

What Burnout Does Whatever the cause, we find that burnout often creates what Wilkins (2000) terms unconditional positive disregard. Wilkins writes, "A distressed, bored or unengaged therapist is in danger of offering unconditional positive disregard" (p. 26). Sometimes counselors in this predicament, though usually unaware, have decided that they just can't hear anymore and certainly can't feel pain with clients any further. Yet, counselors in such predicaments sometimes do not stop work and continue to hold "counseling sessions." Unfortunately, entering sessions in such a state makes engaging clients through UPR nearly impossible.

Thoughts on Avoiding Burnout We may not have great answers for avoiding burnout and hope you will seek other authors who can offer more than the few suggestions we describe here. First, although it is not simple, work to drop your agendas for and assumptions about your clients. Strive to meet each client with UPR, deep empathy, and therapeutic listening.

Second, especially when serving clients with very difficult situations, research the effectiveness of your work. Jeff once agreed to serve a population of students of whom more than half could be expected to drop out of or fail school. Not all of them could be served in counseling due to time limitations. So this situation set up an opportunity for a comparison of the progress of students served in counseling and those unserved in counseling (the two groups were matched in key variables and received alternate services). Our group of educators conducting research found that those unserved continued to drop out or fail at about the same rate, but those receiving counseling came to succeed at a significantly higher rate. These findings were very encouraging within otherwise very discouraging work.

Researching the effectiveness of your work does not have to be complicated. Experimental designs are not even necessary. Qualitative/quantitative studies that indicate how persons served in counseling may have progressed can also be very encouraging. Our point is that sometimes in the face of clients who experience very difficult lives and who probably will not make the progress that we wish for them, it can come to seem like they make no progress at all. Yet, careful but not necessarily complex research can show that they are making significant progress. Such findings can serve to keep counselors going, to keep them open to feeling with clients and ready to accept clients for who they are, and to help them accept the rate of progress that seems possible in their lives.

Third, we encourage you to remember to care for yourself as well as your clients. Strive to base your self-care on being, rather than doing. A void in self-acceptance to be filled only by compiling accomplishments will never be filled. It will become a bottomless pit that no amount of accomplishment will fill.

We encourage you to seek and accept help from others. We find ourselves fortunate in that when we stop to notice, there are others in our lives willing to offer support to us. The question in our self-care is if we will take it or not.

We encourage you to rest and recreate, to do some of the things in life that bring you pleasure, happiness, and joy. Stop and smell the roses.

And, as is important to us and supportive of the UPR that we hold for others, consider renewing or reinvigorating your spiritual faith, if this is at all an interest of yours, and in whatever form it may take for you. It is helpful to us to know that we are not all-powerful as individuals. We believe there is a positive life force, a collective self-actualization process that is vastly more than us as individuals and our works.

Lack of Self-Acceptance

It is difficult to accept others when counselors do not accept themselves. It seems that we humans often criticize or refuse to accept aspects of others that we find unacceptable in ourselves.

However, we do not want to imply that we personally are anywhere close to perfect self-acceptance. We are thankful this does not seem necessary. But as we strive to accept ourselves without conditions, to accept ourselves for being rather than doing, and for being just how we are—mistakes included, this striving seems to encourage and increase our capacity to accept others unconditionally.

Counselors Inadvertently Seeking
to Fill Their Own Needs through Clients

If we, or you, wanted too much for clients to show like or appreciation for their counseling, it would lead to conditional positive regard—accepting clients only as they express that appreciation. Often clients do express appreciation, but counseling can be a thankless as well as low-paying job. We believe it is best for our and your self-care, as well as most efficient in counseling, to have no expectation of being liked or appreciated by clients. Perhaps it is obvious to you, but we find it important for counselors to remember that the counseling is not about them. It is best to get needs for being liked and appreciated met elsewhere.

Another need that a counselor might mistakenly try to have met by clients is a need to see herself as needed. This may set up the conditional positive regard that shapes clients to continue to need their counselor. The inadvertent message from the counselor's behaviors, ways of being and responding, would be, "I'll accept, value, and prize the parts of you that need me (as opposed to all parts of you equally)."

Still another such need is to see one's self as helpful and effective. The message of the counselor's behaviors, ways of being, and responding here would be, "I'll accept and prize you when you change." This would set up the condition of accepting clients only when they obviously change and improve. This condition would inhibit client self-acceptance and thus inhibit change and improvement.

The Analytic Mind

Some counselors have brilliantly analytic minds. When this is the case, such counselors can sometimes come to feel bored when their client is not doing or saying something clearly symbolic or deeply meaningful. If such a counselor is not careful, this can set up a dynamic of conditional acceptance, communicating, "I'll accept you when your actions or works are clearly symbolic or deeply meaningful." We think that early in her work this may have been sometimes true of Nancy. Fortunately, Jeff does not seem to have such a brilliantly analytic mind.

Some Clients Are Hard to Like

Often the abrasive (perhaps surly, or manipulative, or condescending, or annoying) parts of clients come to the forefront in counseling. This is a good thing. As these aspects of your clients meet with your UPR, deep empathy, and therapeutic listening, this initiates your clients' process of deciding just who and how they want to be. An illustrative example from Jeff's work follows.

I once made the mistake of smiling or grinning warmly when a client lit into me angrily for not helping her the way she wanted. The content of her attack seemed to be that if I were a really good counselor, I'd surely be at some more prestigious school than the one that accepted her. Most unfortunately, she misinterpreted my inadvertent smile as amusement with her. So, I had to explain.

What was going on in my mind was that I had known that she was hurting very much and acting in self-hurtful ways. Yet, she had been slow to start the counseling process that I believed would help her. She very much wanted an immediate solution, one that would not require her to engage in a process of self-discovery, that would not require her to begin by showing me who she really was. So, when she finally expressed anger, I saw it as the start of a meaningful journey that I had hoped she would choose.

I did explain my mistaken smile and she did continue. I can't say she came to live happily ever after; perhaps no one does, and her life had been especially difficult. However, she did let me know who she really was, both her surly, sharp parts and her child-like vulnerable parts. And she did achieve a much greater level of self-acceptance, happiness, and success.

I say that my smile was a mistake not just because it was misinterpreted but also because I slipped from empathy with the anger she was expressing to my own personal pleasure at seeing what I took to be progress. My smile also signaled a mistake because it marked my pleasure at her taking the path that was my hope, my agenda for her. I quickly noted my mistake and got back on track.

Clients Doing or Saying Things That Run
Counter to Their Counselor's Moral Constructs

Conflicts between client behavior and counselors' values are inevitable. For example, we have strong moral constructs around the care of children. We believe that good, conscientious, loving care of children is right, and is what all persons raising children should do; and if persons are not ready to do this, they should not have children. Yet, we have had clients who got pregnant, not because they wanted to love and care for children but because they hoped the pregnancy would make permanent a failing romantic relationship, one which they knew was not based on love. We have also had clients who attempted to justify in sessions their failure to pay child support.

It was a great challenge to work through our moral constructs enough to maintain warm acceptance for such clients. It took work, effort, and support for us to maintain our faith that our acceptance would be part of healing for such clients, and so also for their children (while it is rare, we don't mean to say that we would never express our views of clients behavior in counseling; we address that topic in the next chapter).

We have often noticed that we and our students and friends, "get the clients we need." What we mean is that we find that counselors tend to meet the clients that fit right into their weak spots and force them to grow. This is a blessing, we guess. Certainly we have grown from clients that fit into our weak spots. Those persons have brought us opportunities to expand our capacity to accept.

Initial Judging Thoughts

Rogers (1961) explained that it is necessary to "*permit* oneself to understand another" (p. 18). He explained that for himself and many people, our first reactions to others' statements are almost always evaluations. When someone expresses some feeling, attitude, or belief, our base response is usually to think, "That's right" or "That's stupid," "That's normal" or "That's not very nice" (p. 18). Rogers explained that we rarely permit ourselves to understand precisely the meaning of the statement to the other person. He explained that such understanding is risky. If we human individuals understand this other person, it might change us—it might challenge and contradict our beliefs about ourselves and our world in unknown ways. And we humans tend to fear the unknown. So, counselors must avoid the fear of such understanding and change. Outside of awareness, this fear would certainly inhibit UPR and empathy.

HOW UPR IN COUNSELING MAY RELATE
TO UPR OUTSIDE OF COUNSELING

We have noticed that we seem to experience less consistent UPR outside of counseling sessions than in counseling sessions. However, we are pleased that working to maintain UPR in sessions has helped us experience UPR more outside of sessions.

Sometimes the strategic thinking parts of our minds almost believe that UPR might not always be best outside of counseling. Yet, we find that the more we maintain UPR in life, the greater the strategic effect (i.e., the more easily we are able to influence situations, the more others want to support us, the less persons with opposing views want to oppose us). Ironically, if we worked to maintain UPR in order to achieve strategic purposes, that UPR would not be real. UPR is not a means to an end. It is a worthy end, in and of itself.

ACTIVITIES AND RESOURCES
FOR FURTHER STUDY

- Revisit the focus activities of this chapter now that you have more deeply considered the meaning, purpose, importance, and inhibiting factors of UPR. How might you now explain the paradox from Focus Activity 1 differently?

- Regarding Focus Activity 2, journal and discuss your new considerations with a group of peers. How have your answers and related thoughts of UPR changed? How have they been strengthened by what you read? How might you want the impact and fullness of UPR in your relationships to change? How can you work toward this change?

- Contemplate, journal, and discuss the factors that inhibit or that you anticipate will inhibit your expression of UPR in sessions. For example, what are your moral constructs that may make it hard for you to accept certain client behaviors and communications?

- Also contemplate, journal, and discuss how you perceive UPR from others. What makes it easy for you to realize UPR from others? What makes it difficult?

- Journal and discuss what the presence and absence of UPR has meant in your life, and what it may mean and what it may effect as UPR becomes a more consistent presence in your life.

- Now that you have a fuller understanding of UPR, hold practice sessions with partners. Continue your skill practice with therapeutic listening and expressing empathy, but now focus your attention on the levels of UPR that you genuinely maintain. In reviewing your work, consider Wilkins' (2000) clarification of UPR made by reversing each of its elements. Are there evidences of UPR's reversed elements in your work in these mock sessions (e.g., conditional positive regard)? Again, we suggest that the person in the role of the client discuss a topic with some real emotional content. Also, we suggest that you have an observer and/or video tape your session so that you can collect feedback from these sources as well as from the speaker and from your recollection. How well do you think you maintained UPR? How did you express it? (Remember that UPR is expressed more powerfully and purely through ways of being rather than by trying to state it outright.) So what did your ways of being that may have expressed UPR look like? How well and how was your UPR for the speaker perceived? What do some of your strengths and inhibiting factors for holding your client in UPR seem to be?

- Now that you have a greater understanding of UPR, observe, then journal and discuss ways you see that UPR is and is not expressed, and how it is and is not perceived.

- Try to think of alternate ways of defining and explaining what UPR is and why it promotes persons' progress in self-actualization. Journal and discuss the meanings of these alternate ways of defining and explaining UPR in order to develop understandings that fit personally for you.

- We strongly urge you to seek out and read for yourself the sources we have used to develop our explanations of UPR in this chapter, especially works by Carl Rogers and Paul Wilkins' article, "Unconditional Positive Regard Reconsidered," *British Journal of Guidance and Counseling, 28,* no. 1, 2000. UPR is certainly an abstract concept, although it need not be esoteric. Our hope is that you will continue to study it in order to broaden and deepen your understanding of it.

- Another excellent source for further reading and thought on UPR is Jerold Bozarth and Paul Wilkin's book of edited essays that help to further clarify UPR's meaning and importance, exemplify its application, and link it to the wider context of counseling: *Unconditional Positive Regard,* (2002) PCCS Books: Ross-on-Wye, UK.

- Revisit the Primary Skill Objectives of this chapter and verify that you have mastered them to your satisfaction at this time. If not, seek additional practice, reading, and discussion to achieve the level of accomplishment that satisfies you.

- Take time to read and contemplate the following poem, written by our friend, Armin Klein, who has a wealth of rich experience as a psychotherapist and who reflected on aspects of that experience in creating this poem.

Unconditional Positive Regard
Deep Openness

When people come to me asking for help
    With their confusion about themselves, about their lives,
I offer to try to facilitate their own growth process,
    Their own self explorations.

I try to open myself, empty myself of any thinking about them
    Before I meet them. That is very difficult for me.
My background, my early training, has always been to think,
    to problem-solve.
    What knowledge, what identities will provide
      understanding?
At first, my openness—when I can reach it—feels small and
    guarded,
    Scared as I am about new interactions and scared as I am,
      especially,
About facing the unknown—without structures.
    As I try to open, I cannot fool myself that there is any place
To which I am going other than to the unknown and to the
    unpredictable.

Note: We have included the poem above with permission of the author. No further reproduction is authorized without permission from the author.

So I am fearful, though I have grown very comfortable
   with my fears,
    Excited about the growth they promise me.
When we meet, I begin to relax as I begin to sense, to
   experience,
   Something more of the person, the individual being.
Some of this comes from our talk.
    More sensing of their being comes from our non-verbal
       communication,
The moments and the ways in which we each smile, we cry,
   we laugh, and we frown.
    We touch each other powerfully in those moments—
       without words.
We facilitate each other—and my openness grows.
    My mind empties itself of my conditioning—and all of its
       structures.
       My understanding becomes less "problem-solving."
          I experience it, and myself, as more deeply empathic.
       My empathy comes more from my inner self, what I
          like to call my heart.
          I experience myself as more deeply genuine.
    When I become very open, or very "empty"—to the
       extent that I can—
       My interest simply, and apparently naturally, deepens
          and changes.
    My interest becomes very loving—without any apparent
       motivation.
       I experience my positive regard as unconditional.
       I feel myself moving much further into that way of being.

★ ★ ★

I love the process of opening, emptying, that has developed
   for me in my work.
    It has enriched and changed my work, which I love.
This process, however, has also changed my life.
    It has become a wishful model for all my relationships and
       my self exploration,
Although the framework of non-therapy relationships is
   markedly different.

In psychotherapy, I am devoted to the explorations of the
   other person.
    I am always surprised at the unexpected gifts I receive
       from their explorations.
In my friendships, the structure jumps over all the possibilities.
    The responsibilities are more in the background, and they
       are more shared.

I am still, however, always trying to be more open and more
    empty of culture.
    When I succeed, I feel, here also, very loving.
        I experience my positive regard as unconditional.

This model has become my vision of how, with many ups
    and downs,
    I am trying to live. The ups are joyful.
I sense the deepening of my openness and my emptiness as
    Being at the core of the successes and joy
That I have both in my work, and in my friendships.
    When I am less open, closing a little or closing a lot,
        I diminish my genuineness,
The people who work with me in psychotherapy are very
    forgiving.
    My friends are also very forgiving.
I am very grateful and encouraged that both groups of people
    Recognize and treasure my struggle to be deeply open
        with them and myself.

    Reaching for the place of deeper, loving openness brings me
    To Unconditional Positive Regard.

ARMIN KLEIN
January 2001

6

The Delicate Balance of Providing Empathy and UPR in a Genuine Manner

The most exhausting thing in life . . . is being insincere.
ANNE MORROW LINDBERGH

Truth can never be told so as to be understood, and not be believed.
WILLIAM BLAKE

The essence of therapy is embodied in the therapist.
BRUCE E. WAMPOLD

PRIMARY SKILL OBJECTIVES

- Be able to explain the meaning of the term *genuineness in counseling.*
- Be able to explain the meaning of the term *congruence in counseling.*
- Be able to explain common beginning counselor misunderstandings of these terms.
- Understand and be able to explain difficulties that beginning counselors often face in providing empathy and UPR in a genuine manner.
- Discern and be able to discuss the difficulties that you experience or anticipate experiencing in providing empathy and UPR in a genuine manner.

- Understand and be able to explain why your genuineness is important to the positive outcomes of your counseling sessions.
- Understand and be able to explain when and how you would express your thoughts and emotional reactions to clients, as well as how your decision process would play out.

FOCUS ACTIVITY

Observe, take notes for yourself, journal, and discuss ways that persons (yourself and your acquaintances, friends, family) are and are not genuine in everyday communication. For example, do you notice that persons in your life sometimes express understanding, agreement, or concern with statements like, "I understand," "I know what you are going through," "I feel you," or "I feel for you," when they do not understand, know, or feel with you, or at least when these statements are only partially true? In what other ways and situations do you notice persons saying or expressing one thing when it seems that they really mean something different? What seem to be the functions of these mismatches between persons' internal reactions and external expressions? How might these mismatches at times be helpful and at times be hurtful?

For this observation and note taking, we suggest that you keep these questions in mind over the course of a week. It would likely be distracting and perhaps rude to stop to take notes during conversations. So, establish at least two times a day that you stop and reflect on your communications that day and contemplate your and others' genuineness in communication.

Please note that we do not want this exercise to set up criticism of communication that may not be genuine. Such communication may serve a desired purpose or may be evidence of persons coming as close as possible to fully genuine communication in that moment. Communication outside of counseling is very different from communication in counseling. This difference is one of the things that make counseling powerfully therapeutic. Also, completely genuine communication may not even be possible in counseling, although it is an ideal and goal to strive for.

INTRODUCTION

Speaking for ourselves and our friends and loved ones, we know that we are not consistently fully genuine. We, or at least Jeff anyway, used to think that it could sometimes be strategically preferable, politically expedient, to *not* be fully genuine (We hope we didn't just make Jeff sound loose with the truth or manipulative). By now, our experience has taught us that the more open we are to letting others know who we really are and what our experience really is, the better for everyone, the closer we are to others, and the more joy-filled our lives have become. Beyond seeming like the right way to be, this way of being has also proven effective in getting good works accomplished. When we are genuine and respectful with others, we find that others more often respond in kind. Thus, we are more able to find common ground when we disagree and are more able to give and receive support.

Yet genuineness in counseling, like most counseling concepts, is much more complex than it sounds. For an overly simple thought to spur your contemplations, please consider: As difficult as it is to have one person fully understand another, if a counselor tried to have a client fully understand his experience of this client (i.e., the counselor's thoughts about this client, feelings in reaction to this client), the explanation would take so long that the roles would switch. The counselor would become the client or primary speaker explaining his experience, and the client would become the listener. Therefore, counselors must carefully choose what limited parts of their experience of clients to communicate. Further, we titled this chapter "The Delicate Balance of Providing Empathy and UPR in a Genuine Manner" because we want you to focus much more on a genuine expression of your empathy and UPR than on a genuine expression of your whole self and your full set of reactions to clients. We want you to realize the delicate balance in decision making between remaining focused on your client and sharing your experience of your client.

WHAT GENUINENESS MEANS AND DOES NOT MEAN

As with each of the core conditions of counseling, Rogers (1957, 1961, 1980) can be credited with bringing the importance of counselor genuineness to the full consideration of the helping professions. Rogers (1980) explained that this core condition or element of the counselor's way of being could be called "genuineness, realness, or congruence" (p. 115). He explained, "The more the therapist is himself or herself in the relationship, putting up no professional front or personal façade, the greater is the likelihood that the client will change and grow in a constructive manner" (1980, p.115). So to us this means that the counselor cannot hide behind the façade of being the assessor (i.e., taking on a role of assessing or analyzing her client, rather than meeting her client fully and experiencing with her client). It means that the counselor cannot hide behind a professional persona of being all-knowing and omnipotent, of probing and prescribing what her clients should communicate and experience in their counseling sessions (and fortunately, being all-knowing and omnipotent is not necessary).

Rogers explained that the counselor attitude of genuineness means that the counselor "is openly being the feelings and attitudes that are flowing within at the moment" (1980, p. 115). He explained that for the counselor, "what he or she is experiencing is available to awareness, can be lived in the relationship, and can be communicated, if appropriate" (p. 115). In our section, "How Counselor Genuineness Is and Is Not Communicated," we lead you through an exploration of when it may and may not be appropriate for you to communicate your experiences of your client that occur outside of empathy and UPR to your client. For this beginning section, please know that by genuineness, we do not mean communicating any or all of your experiences in sessions with your clients to your clients. If a counselor tried to communicate all of her thoughts and feelings in a session to a client, some of the information would overwhelm the client and inadvertently shift the session's focus from the client's communication. We want you to think of genuineness as you making your strivings for empathy with and UPR for your clients every bit as much of who you are as possible. It would not be very helpful for you to just be an empathy and UPR machine. While in some moments it may be the best a counselor can do, it would not be effective for you to go through the motions of empathy and UPR that you do not really feel.

Rogers explained that the core condition of genuineness means "the therapist makes himself or herself transparent to the client; the client can see right through what the therapist *is* in the relationship; the client experiences no holding back on the part of the therapist" (1980, p. 115). This notion of your clients experiencing no holding back on your part underscores the importance of your working through the things that get in the way of your experiencing empathy and UPR. For example, we believe that all people are at least somewhat inhibited from full experience of emotions. However, if your inhibitions of full emotional experience significantly inhibit you from experiencing your clients' emotions, this may cause you to go through the motions or say words as if you are fully experiencing with your client, when in truth you are holding yourself back. Your work might then look like counseling on the surface but not actually be therapeutic. At some level, clients know when their counselor is holding back in this way, when their counselor is merely putting up a façade of empathy. Such a façade is not true empathy and *will* limit client progress.

An additional quote from Rogers (1980) may clarify: "To be with another in this way [deep empathy and UPR] means that for the time being, you lay aside your own views and values in order to enter another's world without prejudice" (p. 143). We know that laying aside our views and values can be difficult, and we expect that may be true of you, too. It may be an impossible goal to reach in perfection, but it is essential to your clients' progress that you approach this goal, that you come close to it most of the time, and that you continually work to expand your capacity for empathy and UPR so that you continually come closer to this goal in perfection.

A SAMPLE OF LITERATURE SUPPORTING AND CLARIFYING GENUINENESS AND ITS IMPORTANCE

We suspect that it is even more difficult to research the outcome effects of genuineness or congruence than empathy and UPR because, as we see it, genuineness is a way of being that underlies providing true empathy and UPR, rather than a separate quality or set of behaviors. However, in this section we suggest a small sample of sources supporting and clarifying the importance of genuineness.

Understandably, most outcome research into genuineness or congruence has focused on congruence as part of the set of core conditions of therapeutic interactions. Patterson (1984) explained that such studies provide "a body of research that is among the largest for any topic of similar size in the field of psychology" (p. 431). Fortunately, numerous meta-analytic reviews can help us sort this body of research. Truax and Carkhuff (1967) asserted that the core conditions were effective with a wide range of clients and among counselors, regardless of training or theoretical orientation. Consider their concluding statement:

> [C]ounselors who are accurately empathic, non-possessively warm in attitude, and genuine are indeed effective. The greater the degree to which these elements are present in the therapeutic encounter, the greater is the resulting constructive personality change in the patient (p. 100).

Truax and Mitchell (1971) offered similar conclusions, explaining that empathy and warmth can only be facilitative when they are at least minimally real in the counselor. Orlinsky and Howard (1978) reported that two-thirds of the studies they reviewed on genuineness (20 in all) showed a significant positive relationship of genuineness to outcome. In their later review (1986), Orlinsky and Howard reported that 20 of 53 studies showed the same positive relationship and noted that this included studies that used client perception as part of the assessment of counselor genuineness. However, it seems fair to note that other authors have suggested methodological limitations of conclusions in support of genuineness (Cramer, 1990; Lockhart, 1984; Watson, 1984).

Numerous studies have used measures of the match or mismatch between counselors' expressed message and messages that are implied through the way things are said or through nonverbal behavior as a method of investigating the relationship between genuineness and predictable positive outcome. Findings by Haase and Tepper (1972) suggest that congruence is an essential underpinning of the communication of empathy and that inconsistency between verbal and nonverbal counselor messages undermine even highly empathic messages. Findings by Graves and Robinson (1976) suggest that inconsistent counselor verbal and nonverbal messages may prompt clients to see them as less genuine and prompt clients to maintain greater interpersonal distances from their counselor, especially when nonverbal messages are negative and verbal messages positive. Further, from this study it was suggested that when verbal and nonverbal messages have equal intensity, the credibility of the counselor's message is enhanced. Findings by Sherer and Rogers (1980) suggest that counselor nonverbal behavior has an impact on perceptions of counselor warmth, empathy, and genuineness, and that because ratings of warmth, empathy, and genuineness were highly interrelated, the concepts are probably quite similar at a nonverbal level. Findings by Tyson and Wall (1983) seem to concur, suggesting that congruent verbal and nonverbal behavior enhances the impact of the counselor's message.

Grafanaki and McLeod (1995) provided research suggesting that counselor congruence is an important part of helpful events in sessions and counselor incongruence is an important part of hurtful events. However, relationships between congruence or genuineness and positive outcome are complex. For example, Grafanaki (2001) gives the following example from this research:

> [T]here were times that the counsellor was congruent and was trying to "direct the therapy process" or "focus the discussion on an important issue," however, these efforts did not always have a positive therapeutic impact. It appeared that the client's readiness to work on a particular issue determined the helpfulness of the counsellor's response rather than whether the counsellor was congruent or not.

THE IMPORTANCE OF GENUINENESS IN COUNSELING

Keeping Therapeutic Listening, Empathy, and UPR Real: A Therapeutic Relationship with a Real Person

Making the therapeutic listening, empathy, and UPR that you provide your clients an integrated part of who you are means that their therapeutic relationship is with a real person. It means these skills and ways of being are not a façade that covers you but are real and really are who you are with your clients. In Chapter 3, we explained part of the importance of empathy through our conceptualization of what drives conduct-disordered (CD) behaviors in troubled children (Cochran & Cochran, 1999). We assert that therapeutic relationships focused on therapeutic listening, empathy, and UPR break down belief systems that may drive CD behavior, such as believing "I am unlovable and no one can understand me. Further, I can't stand the inevitable rejection. So, I must hurt others to drive them away from me now." However, if you were just going through the motions of therapeutic listening, empathy, and UPR, then your client with CD behaviors would not be experiencing a relationship with a counselor who is real and would not have the opportunity of a therapeutic relationship providing real evidence that his troubling belief system may not be true. If, instead of engaging in a therapeutic relationship, a client spoke into a tape recorder and then listened to what he said on the tape, this might have some minimal therapeutic value. However, that experience would be vastly less than experiencing a therapeutic relationship with a counselor who has integrated therapeutic listening, empathy, and UPR into her genuine way of being with clients.

The Connection to and Role of Genuineness in the Set of Core Conditions

Genuineness and Empathy Bozarth (1999) describes a "conditions loop" (p. 80) in which the three core conditions are ultimately one. Bozarth explains a loop of genuineness with empathy, in that the more a therapist is able to be aware of her own experience, the more that therapist will be able to be aware of her client's experience. Then, the more that therapist is able to be aware of her client's experience, the more she is able to be aware of her own experience. Thus, Bozarth defines genuineness as "a natural awareness of the therapist to her own experience" (p. 80).

We take Bozarth's explanation of a conditions loop and his definition of genuineness to mean that ideally you would be fully aware of your client's experience in sessions and secondarily aware of your own experience in sessions. To us, the importance of your self-awareness in sessions is not so that this experience can be expressed to clients but so that you can manage it, so that it enhances rather than detracts from your therapeutic listening, empathy, and UPR. Satir (2000) wrote of her work:

> When I am in touch with myself, my feelings, my thoughts, with what I see and hear, I am growing toward becoming a more integrated self. I am more congruent, I am more "whole," and I am able to make greater contact with the other person. (p. 24)

Genuineness and UPR Lietaer (1984) views congruence or genuineness, and unconditionality or UPR as parts of a more basic attitude: "openness." Lietaer asserts that openness with self creates a greater openness with clients.

> The more I accept myself and am able to be present in a comfortable way with everything that bubbles up in me, without fear or defense, the more I can be receptive to everything that lives in my client. (p. 44)

Modeling

We see counseling as a process of helping clients become congruent. Therapeutic listening, empathy, UPR, and genuineness play roles in this process. Therapeutic listening and empathy help clients realize who they are and who they want to be. UPR provides a safe and nurturing environment in which clients allow their counselor and themselves to know who they really are and who they want to be. Genuineness makes empathy and UPR real, rather than techniques, but also provides clients with a model of a congruent helper and person.

While we often refer to the three core conditions, Rogers (1957) actually introduced six core conditions for therapeutic personality change. One is that the persons be in psychological contact. Two more are that the therapist experiences UPR and empathic understanding. Another is that this empathy and UPR are at least minimally communicated. The other two include the client being in a state of incongruence and the therapist being in a state of congruence.

We believe the therapist's state of congruence, with empathy and UPR integrated into her being during her time with clients, has a powerful modeling effect. The conditions for modeling are present. The two persons have a close, personal relationship. The first person tends to be vulnerable through incongruence, through the nature of seeking or needing help, and through the fact of entering a world that is new to him but familiar to his counselor. So clients tend to look up to and respect their counselors, even if they are frustrated or cross with them at times. Thus, the counselor's modeling of genuine empathy and UPR helps clients come to have self-empathy, to fully realize what they feel, and to develop self-UPR, to fully accept who they are even while they are motivated to improve.

Jeff has had a number of his adult clients illustrate this process for him in a way that may help you understand.

> Some of my particularly articulate adult clients have told me at the end of our work together that early in our counseling, when faced with a painful choice or dilemma, they would hear my voice in their head fully accepting and understanding what they were going through. They told me that they used this voice to make decisions full with care for themselves and others. Later, this voice became their voice, and they came to realize that they were coming to need time with me much less.

> When I first heard this from a client during our therapeutic process, I was alarmed that I was being overly influential, perhaps inadvertently shaping my clients to be like me. However, I have come to see this as a process of clients modeling their treatment of self and others after the empathy and UPR they experienced from me. While I have been honored and privileged to have this role, I have come to know that I was not influencing my clients to be like me, especially when I realized that they made decisions, with full self-empathy and UPR, that were best for them but that were not always the decisions that I believe I would have made in their place or that I would have suggested for them, had suggestions been a part of my role.

Creating a Safe Place for Emotional Honesty

Counselor congruence allows counselors to be perceived as safe and solid to clients. If a counselor goes through the motions of empathy and UPR but these motions are not genuine, clients can sense this incongruence at some level. Then, having a vague awareness that something is untrue in their counselor, the clients understandably become suspicious of their counselor and see their counselor's actions as only tricks or techniques.

A metaphor may be helpful in understanding this effect. We, and many people we know, have experienced romantic or other relationships where one person wants it to work out so badly that he tries to make himself into just the person that he believes his partner wants. That person is then in a state of incongruence. When this happens, the partners do not share a relationship between two solid persons but a relationship between one person who may be solid and one person who clearly is not. Such relationships usually dissolve into mistrust. Such relationships often leave the partner who may be solid wondering, "Just who are you?" and thinking in frustration, "How can I find out how I fit with you, when I can't tell who you really are?"

Another example of incongruence may help further understanding. We have noticed that some adults will ask children to fully disclose on a topic, even when the adult is withholding key information. Jeff observed a blatant example of this in an elementary school.

> A school adult pulled a child into a private meeting to discuss an incident of misbehavior on the bus. The adult had already decided that she knew what this child had done, the details of what happened, and who was at fault. Yet, she opened the discussion of bus behavior by asking the child just what had happened on the bus with feigned open-mindedness. Understandably, the child clammed up and became unwilling to talk about it. After the child left, the adult complained, "How can I help him, if he won't be honest with me?" I saw this interchange as a card game in which the dealer insists that the other player show all his cards and tell his strategy, while this dealer continues to cover her cards and hide her strategy, then complains that the other player is unwilling to play by these rules.

HOW COUNSELOR GENUINENESS
IS AND IS NOT COMMUNICATED

Declarations of Genuineness Are Rarely Helpful

Your genuineness is not something you can express in a statement. Rather, it is a way of being to strive for. When your empathy and UPR are genuine, they will come close to being consistent in you. Just as the child in the bus incident knew not to trust the terms of the discussion, your clients will usually know when you are being genuine with them and when you are not. If your client continues to doubt your genuineness, it will be helpful to respond to this doubt with therapeutic listening and empathy so that your client can fully realize and learn from this doubt.

Jeff often remembers the futility of claiming genuineness that is not true by remembering that when other teens that he grew up with were exaggerating a story, they would preface the exaggeration by declaring, "Ain't no lie, man." Such statements became a communications flag for Jeff, from which he would note, "The following is probably a distortion of truth, at best."

Sometimes Counselors' Experiences of Clients
"Bubble up" or Cannot Be Hidden

Bozarth (1998) clarifies his definition of genuineness, explaining that genuineness may be purely internal awareness or may involve a statement to a client. Regarding when or how genuineness may be a statement, Bozarth relays Rogers' explanation that when he is really empathic, attuned to his clients, and working with them within their frame of reference, sometimes his statements of his genuine experience of them seem to "bubble up" in him, and he makes the expression out loud. Consider Jeff's description of how this has happened in his work.

Many times I have been surprised by a change in or statement by a client. Other times I have been frustrated by clients' ongoing pain or slip back into self-hurtful patterns that they and I had hoped were over for them. At yet other times, I've been very happy as clients made progress that they had long wanted. I don't have much of a poker face. So, when I have had these feelings in response to clients, my clients have easily known this. Then, when my therapeutic listening and empathy help me realize that my clients have noticed and reacted to my reactions to them, I express my empathy and understanding of their reaction to my reaction.

For example, when a client noticed that I seemed frustrated as he told me of slipping back into a pattern of depression and self-doubt after having made much progress, I responded, "You seemed to see a part of my reaction cross my face, and it seemed that a look of alarm crossed yours." He then half-heartedly disclaimed his reaction, saying, "Oh no, it was nothing," then looked down in silence. To which I tenderly responded, "You say it was nothing, but now you seem intimidated, hurt by me." Then, he let more of what I thought was his true feeling show, blurting, "Well you're probably all critical of me, thinking I just make my own problems." Realizing his concern, I was glad to reflect and explain my reaction, saying, "Oh, that is what you figured was going on with me. No, I am not thinking critically of you, but I do feel frustrated with this set-back. I know how much you have wanted to feel better." So he told me, "Yeah, I'm frustrated, too." Then, I of course shifted myself back to therapeutic listening, empathy, and UPR. I responded, "So, you too. I also have the thought, that I'm not sure of but I got the impression a moment ago, that *you* are critical of you, and you have the thought that you make your own problems." This then seemed to open a door for him to come to realize that while it was true that he very much wanted to feel better, his depression was also a temptation and served a purpose of his avoiding responsibility. A representative thought pattern for this might be, "I can't help it, I'm Depressed." In such a thought pattern, Depression gets a capital D, as if it is a defining label, rather than a description of a set of symptoms. My expression of frustration was an accident, but within the context of careful listening, empathy, and UPR, it became very useful to this client's self-understanding, to his understanding of how I saw him, and of course to his progress.

State Your Reaction When Your Reaction to Clients Interferes with Your Empathy and UPR

Stating enough of your reaction for your client to understand what is going on with you is sometimes necessary to stay on track with a relationship focused in empathy and UPR. Consider the following examples from Jeff's work to illustrate our point and show how this may happen.

Example 1 I got the news of the unexpected death of a friend's child just before beginning a small group session. There was nothing I could do to help my friend immediately, so I kept my session appointment. As more than one client was telling the group and me how they felt overwhelmed by stresses of school and work, I had begun to listen poorly, but I didn't realize it right away. So, without realizing what I was doing, I phrased a reflection something like this: "So, these little things are really starting to seem overwhelming to you." Fortunately, a group member awoke me from my lack of awareness by immediately calling me on my error, saying, "These things aren't little to us." As I attempted to get back on track and reflect correctly, she added, "And you seem bored today, like you don't think these things are that important." To which I just told the truth, something like, "Yes, I am having difficulty focusing and listening right now." I added explanation because I wanted her and the others to understand that my error was due to factors outside of the group, saying, "I just got the news of the death of a friend's child before we met, and I'm having difficulty getting the pain that I imagine my friend feels out of my mind so that I can focus with you, here." Having said that, I became more able to focus on group members' communications, reactions to me, and to each other. The group members soon moved on and continued to share the stressors that were important to them and also seemed to come to see their stressors in the context of the news that I shared with them. Again, my lapse in empathy was an accident that forced an explanation of my incongruence. That explanation may have been mildly helpful for group members to scale the stressors they were experiencing. However, that small outcome was only an accident. I don't think the group members perceived their stressors in any particularly distorted way before my lapse in congruent empathy and following explanation.

Example 2 In another example, I was working with a client whom I had begun to think was trying to get me to dislike him by making racist and political statements that he seemed to assume I strongly disagreed with. Through these statements over a few sessions, I worked to keep my focus on his experience, to lay aside my views and values in order to enter his world without prejudice. For example, he might have said as a side note to his worries over not getting hired by a top firm, "Well, you probably don't agree with this, but the blacks and women are using the courts to every advantage they can get." While I did feel an urge to argue with this side statement, I decided that the crux of what he was expressing was his fear that he might not get the job he had dreamed of. So this was what I focused my reflection on: "You have this anger toward blacks and women and courts, and your great worry is that you'll never get the job with _____." However, when he later told me that he didn't pay his child support because his ex-wife manipulated the courts and other events against him, I knew that his views and actions had so affected me that I was listening poorly. So, I decided to tell him the truth of my reaction to him. I stopped him and explained, "I am having a hard time listening and understanding your experience, as I think I strongly disagree with some of your views." I then gave a couple of examples of my disagreement, that I thought child support was owed to the child, regardless of the other parent's behavior; and that I thought it unfair to blame blacks and women for whether or not he was able to get the job he dreamed of. I then managed a warm process reflection, saying, "Also, I have the idea that you've made these statements in shocking ways, assuming I would react critically." He explained that he'd not intended to shock me and that these were ideas that he firmly held, and he did figure I strongly disagreed. Yet, he also seemed to consider my process reflection as at least possible. While we continued to work together for some time after that moment, we certainly never argued about his views. Additionally, in the time that we were together, he began sending his child support, and when he expressed views that he assumed I would disagree with, he stated them as his opinion, rather than fact. It seems to me that his views softened over the time we had together. Certainly, he became much more open with his emotions, expressing them as his emotions more often than as political views. However, I attribute his change much less to that moment when I expressed my views in disagreement than to the many more moments when I successfully responded to him with empathy and UPR.

Please realize that it would be a problem for us or any counselor if the numbers of statements or persons to whom we have difficulty listening with full empathy and UPR were significant. More important than expressing your disagreement in rare moments when your listening is greatly inhibited is ever expanding your capacity for empathy and UPR with greater ranges of persons and topics. This is our ongoing path, as well.

Being Who You Are in the Phrasing of Your Reflections

Your reflections and expressions of empathy are meant to express your understanding or feeling with your clients. Yet, they also are expressions of who you are, what is important to you, and how you see life. You may remember that in Chapter 2 we described each communication from a client as coming to you through the mental filter of your views and that any response from you must pass back through that mental filter. Your understanding of your clients is never a true understanding but the best understanding you can muster, given the limitations of perception and language. This is doubly true for the statements of understanding that you make to your clients. So, while we would like your and our reflections to be true images of clients, they are not. They are a combination of your clients, your striving to understand them, and yourself.

For example, we tend to think a lot and value the existential notions of choice, personal power, and decision making. This is part of the mental filter through which we hear our clients' communications and through which we express our understandings. So, the phrasings of our reflections often are peppered with words that suggest such notions: "You have decided that is not possible for you," or "You are choosing not to do that anymore."

Additionally, significant influences on Jeff's development as a counselor and person come from Rational Emotive Therapy (Ellis, 1989) and Cognitive Behavioral Therapy (Beck & Weishaar, 1989). Belief systems from these theories have become a part of his worldview. Thus, he tends to hear the thought patterns behind his clients' feelings and actions and include these in his reflections. For a different example, a counselor with strong beliefs aligned with feminism might have a mental filter that prompts her to hear client communication in the context of power differentials in clients' lives and the effects of these differentials.

We believe it should never be a counselor's intent to have his worldview affect what he hears and how he responds. However, it will probably happen even when a counselor's intent is to really see the world as it is to his clients. So, each counselor's reflections do become, at least inadvertently, a manifestation of that counselor's worldview. This seems to us yet another important reason for counselors to continually strive to know themselves, to improve themselves, and to ever expand their capacities for understanding, empathizing with, and accepting others.

Make Only Careful, Judicious Self-Expressions, Beyond Your Ever-Present Empathy and UPR

Remember, whether through the phrasings of your reflections or outright expressions of reactions to clients, keep your influence limited, giving favor to helping your client find her own meanings for her experiences. There is a tremendous power differential between client and counselor, in favor of the counselor. By virtue of presenting as clients, your clients are in a state of incongruence and hence are vulnerable. To consider that there might be an equal exchange of views between counselor and client seems *extremely* unlikely to us.

WHAT MAKES THE DELICATE BALANCE OF PROVIDING EMPATHY AND UPR IN A GENUINE MANNER DIFFICULT

The Errant Thought—I Am Who I Am

We have known some beginning counselors to misunderstand the concept of genuineness. Their misunderstanding was that it meant being themselves as they are in whatever current stage of development they have achieved. For example, more than one beginning counselor has expressed to us her resistance to empathy and client process in the following ways: "But I'm a take-charge person. So, when I guide my clients' lives, I'm just being genuine." Or "Look, I'm not a patient person. If I started a long novel, I'd skip the middle to get to the end and see how it comes out. So, I'm being genuine in just skipping my clients to the end."

As you may guess by now, we see a number of errors in such beginner counselor statements. For one thing, such statements seem to assume that the persons making the statements are static, that whatever qualities they have at the time of the statement are the predominant qualities they will *always* have. We know that we and seemingly every person we know are constantly developing. Therefore, we do not accept that any one stage in a person's development is permanent.

However, our key disagreement for this context is over the meaning of genuineness. We urge you to make the core conditions of counseling and their resulting skills the core of who you are in your counseling relationships, instead of trying to somehow shape your way of being in counseling to fit your perception of who you are and what is possible for you in your current stage of development.

The Challenge of Clients Who Are Hard to Like

Clients who are hard to like present a particular challenge and opportunity for counselors to expand their capacities for making therapeutic listening, empathy, and UPR a greater part of who they are in their therapeutic relationships. Whenever we find a client hard to like, it is because that client strikes against some cognitive construct of ours, probably regarding how people should be or act (We gave examples of this in Chapter 5). In such situations, counselors are presented with opportunities to review and perhaps change such constructs in order to be able to accept and empathize with a broader range of client experiences. Supervision or the counselor's own use of therapy offer venues for this work.

As in the example from Jeff's work in the previous section, when you have tried to work through dislike but it persists to the point of inhibiting your therapeutic listening, empathy, and UPR, we think it can be best to say so in as open a way as possible, then to respond with careful therapeutic listening, empathy, and UPR to whatever response your client makes to your statements. On a side topic, please note that we do not believe such statements would be within reason if counseling children. With children, the power differential is far too great, and such statements might also require adult explanations that are above what a child could or should be ready to understand.

The Need for a High Level of Self-Development
for This Counseling Skill

Counselor Maturity and Personal Development   In light of the previous section, it makes perfect sense that Lietaer (1993) points out that congruence requires therapists to be psychologically well developed and integrated. Lietaer goes a step further to assert that personal maturity can be considered a therapist's main instrument in therapy.

Self-Awareness We, of course, see self-awareness as a key to counselor maturity and personal development. Especially, we see it as key to congruence. It would be difficult for a counselor to maintain reasonably consistent congruence if he is not reasonably aware of his internal experiences and external manifestations.

Additionally, reactions to clients are probably unavoidable by counselors, who are, of course, living, feeling, and thinking humans. With insufficient self-awareness, counselors may express reactions of which they are unaware, and may even deny them when asked. Such a denial may prompt client self-doubt and be unknowingly dishonest, modeling just the opposite of the open communication that you would like to see from your clients. Please consider the following explanation by Tudor and Worrall (1994).

> It is likely that if as therapist we consistently ignore or deny some of our feelings and experiences we will, out of awareness, communicate such unassimilated, or partially accommodated, material to our clients. Clients, for instance, who are particularly sensitive to anger, may sense in us an irritation, which may or may not have anything to do with them. If they question us about their sense that we are irritated, we need to be sufficiently aware of our selves that we can identify and acknowledge any grain of truth to their perception if there is one. (p. 2)

Self-Acceptance and Self-Honesty It seems to us that self-acceptance and self-honesty are prerequisite to the high level of self-awareness just described.. If we counselors are highly critical of our experiences, especially our experiences of clients, it will be difficult to see them honestly and clearly. Without this honest, clear sight, brought about through self-acceptance, our self-awareness and any communication of our experience would be seriously inhibited. We think, however, that while we emphasize this high standard for your development, it is also important to remind you that perfection is not human, not possible, and maybe not even desirable. We want you to work hard to move yourself toward this important goal but not to let that work become an inhibiting factor.

Knowing the Difference between Your Reactions to Clients, Your Reactions to Things Other than Your Clients, and Your Shoulds In discussing guidelines for counselors' expressions of their reactions to clients, Tudor and Worrall offered the following explanation, "We need to be able to tell what, of all that we are experiencing as we work, is a response to our client, and what is a response to a bad night's sleep or the morning's post or some similarity between our client and some other person we know" (1994, p. 4). We add that we counselors also need to have worked through enough of our shoulds (i.e., rules that we humans believe that we and/or others should, ought, or must live by, which can sometimes be beneath our awareness; Ellis, 1989) so that these shoulds do not greatly affect our experience of clients. If we haven't worked through them, which can be a lifelong process, we need at least to have brought them enough into our awareness that we are not inadvertently commingling and communicating "reactions to clients" with our shoulds, and using them as a way to covertly shape clients and minimize clients' opportunities to learn from their full experience.

Consider the following example: If you have a should that says, "People should be nice to others and if persons are not nice to others, this is terrible, awful!" you may react to a client who is telling you about things he has done that were hurtful to others by stating your hurt, irritation, or disappointment over his hurtful actions. You might simply tune out from listening and attending during the parts of his sessions in which he is trying to tell you of his hurtful actions. With either of these reactions, you stifle your client's opportunity to learn from fully experiencing his actions through telling them to you and receiving your accepting, empathic response. Having his full experience enhanced by your empathic response can help this client come to realize his dissatisfaction with his actions toward others, then decide to change and own that decision. Further, through your empathic response, he may even learn what the personal meaning of his hurtful actions have been and thus learn how else to find this same meaning. So, we encourage you to continue to work through your shoulds as a way to expand your capacity for consistent therapeutic listening, empathy, and UPR, as a part of who you truly are with your clients.

Still, we counselors are never going to be perfect and free of shoulds. This complete freedom from shoulds may not even be desirable. So, when your clients' actions or communications run so strongly against one of your shoulds that you realize it has distracted you from therapeutic listening, empathy, and UPR, and when that distraction persists despite your best efforts, we suggest you express this interaction of your client's behavior and your belief system openly (as in Jeff's earlier example), rather than allowing your reaction to be a covert, insipient influence over your therapeutic relationship. Then, outside of your sessions, get back to work on reducing the influence that particular should has on your relationships with others and yourself.

The Question of Expressing Your Positive and Negative Experiences of Clients

We have sometimes noticed that beginning counselors are excited to express reactions to clients that could be thought of as positive, such as their admiration of clients surviving adversity or their joy in client progress. We are not saying this is always wrong. However, we believe it important for beginning counselors to realize that to be consistent, the flip side of these positive expressions might be expressing annoyance with clients who seem to have it easy or frustration with a client's lack of clear progress. There would be an equal rationale for expressing these negative reactions. We think that if you choose to express your reactions at all, it would be best to express them consistently, whether positive or negative. That way your clients will know that while your intention is to help them process their experience, when you find it inhibitingly difficult to avoid expressing your experience, you will be honest with it. The contrast to this would be using your reactions to covertly shape clients in a particular direction or to favor ways of being that are pleasing to you over those that are not.

The Need for High-Level Observational Skills, Therapeutic Listening, and Empathy

Sometimes an initial tip that one of us is expressing some subtle incongruence as counselor to our client is the client's reaction to the one of us serving as her counselor. For example, if a client tells of her actions, while moving in a way that suggests wincing, we could guess that she is assuming a criticism from her counselor. If such an assumption of criticism is true, then she might understandably not be fully disclosing. If she were correct in assuming her counselor's criticism, it would be better for the counselor, after reflecting, to either acknowledge the truth of the assumption (in the context of owning his personal reaction—see Scenario A) or honestly dispel the assumption. Either way, such situations require high levels of observational skills, therapeutic listening, and empathy from counselors. The counselor cannot genuinely respond to this client's perception of counselor criticism if he does not sense her perception of criticism. Fortunately, such perception is developed through hard work and attention to developing strong empathy. A dialog illustration of how such a scenario might play out follows. We offer A and B scenarios, based on different counselor internal reactions.

Scenario A

   Counselor: [Picking up after the client had developed a pattern of making a motion that suggests wincing, while telling of her actions.] I notice you almost wince when telling me that. I'm guessing that you're worried I'm thinking critically of you.

   Client: [shrugging] Well, I guess anybody would be.

   Counselor: So, this is an assumption you'd have of anybody, and that includes me.

   Client: [seeming to get irritated] Well come on, everybody knows it's wrong!

   Counselor: [who does have critical thoughts in this instance]: Yes, I do find these actions that you are telling me of hard to listen to. I find myself distancing myself from you when I hear them.

   Client: [After stunned pause, looks at counselor and shrugs.]

   Counselor: While you expected it, you seemed stunned by what I said.

   Client: Yes, I hurt very much over how my actions have affected others. Really, I want to be liked, too.

Scenario B. (picking up at the point that it differs from Scenario A)

Counselor: [who does not have critical thoughts in this instance]: I wasn't thinking in terms of right or wrong in that moment. I mostly was aware of how much you are struggling with this. And now you seem to tell me that part of that struggle is that you see that everybody knows it's wrong.

Client: Yeah, everybody [thoughtful pause]. Oh, I'm so embarrassed.

Counselor: You're embarrassed that so many might have seen your mistakes.

Client: Yeah, I guess I'm embarrassed because I think it's wrong, too.

Clients Who Ask for Your Experience of Them

Some clients ask how you see them. In such situations with adult clients, we are generally glad they ask and are willing to give an honest answer. Such answers can be very high-level reflections. Still, there are some tricky aspects to answering such questions. Because our focus is always intended to be on helping clients fully experience their self and their situation, we would first reflect what clients seem to have communicated with the question. For example, if a client asks one of us, "Am I normal?" depending on the intent of the question that we inferred from our client's tone and the context of her question, we would reflect something like, "That's something you'd like to figure out, and you're thinking it'd be helpful to know my opinion." However, after such reflections, we find that while we are ready to answer the question, our clients often withdraw the question, following instead the trail of their experience in other directions.

But this question, "Am I normal?" brings up other important complications. If your client asks if you think she is normal, that is not really just a question about how you see her. It is also a question that requires knowing what normal is. We don't believe we know what normal is. So, we'll also suggest that you not try to answer that question.

This issue of normalcy also reminds us that if you tell clients how you see them, you should stay away from judgments (i.e., good vs. bad, normal vs. abnormal). In fact, it is best to respond with how you feel in response to your client, rather than how you think. Consider the following answer scenarios.

Scenario A: Client Withdraws the Question

Client: I just can't believe how nervous I am about this. Is that normal?

Counselor: In your shock over how nervous you feel, you wonder if it's even normal to feel so nervous.

Client: Yeah. Of course, I guess I worry over what's normal a lot and that leaves me even more nervous [goes on to express frustration with worry].

Scenario B: Client Does Not Withdraw the Question

(Continuing from counselor reflection of the intent communicated in question)

> **Client:** Yeah. I sure wish I could just know that. What do you think?
>
> **Counselor:** [Speaking slowly, thoughtfully, carefully.] I don't know. I don't think I can figure out what normal is. I try not to think of you in terms of normal and abnormal. [seeing a frustrated look on client's face] You were hoping for more of an answer from me.
>
> **Client:** Yeah. Can't you tell me anything?
>
> **Counselor:** [Saying what he thinks of client without judgments.] I see that what's normal and how others see you is very important to you. Also, that the effects of your decision on you and your family are very important to you. Then, placing value on both these things leaves you feeling very, very stressed and nervous.
>
> **Client:** I think maybe I worry too much about the future. I want to start doing what feels right, right now.

Scenario C: When a Feeling Response Is More Fitting Than Thoughts from the Counselor.

(After having gotten frustrated and yelling at her counselor that if he were really good he'd not be working at this lousy school anyway.)

> **Client:** [Looking both still mad and sheepish at the same time.] Oh God, I bet you hate me.
>
> **Counselor:** [Speaking with a tone of interest and acceptance that was invigorated by her strong emotion.] I would like you to know how I feel toward you, but first I want to note that you expressed strong anger to me that I assume had been building up for some time.
>
> **Client:** [Still frustrated but also calming down.] Oh, I'm just mad today. [pause] But, God, that was mean. You must hate me now.
>
> **Counselor:** You are troubled that I might hate you now. I don't hate you. I know that your situation now is *very* hard for you. I wish it could just be over and you could already be at the place where you've done the work you need and are feeling better. I long for this. But, as part of that work, I'm glad you will show me how you really feel, even if what you feel is anger at me.
>
> **Client:** [Tearful] It's so hard. [short pause] God, this sucks.

A few more notes about these scenarios: If the normalcy question had been about use of counseling, we might have taken more of an opportunity to educate about counseling (see Chapter 8 for examples). Note from Scenario C that while the counselor expresses feelings in response to his client, feeling words are only part of expressing emotion. Feelings are often expressed in tone and wishes, as well as single words for emotions. Additionally, we mean the scenarios to assume work with adults or adolescents. With children, we only find it useful to answer questions about the structure of counseling or questions that are simple and peripheral (i.e., Did you see that show last night?)

The Need to Balance Freedom That Optimizes Personal Connections and Allows Experiences to Bubble Up into Expression with Avoiding Influence That Limits Client Expression

In the classic film demonstration (Shostrom, 1965), the client Gloria finishes an honest and humble statement to Rogers by saying, "All of a sudden while I was talking to you I thought, 'Gee, how nice I can talk to you and I want you to approve of me and I respect you, but I miss that my father couldn't talk to me like you are.' I mean, I'd like to say, 'Gee, I'd like you for my father.' I don't even know why that came to me." Rogers responded to her instantly and warmly, "You look like you'd make a pretty nice daughter. But you really do miss the fact that you couldn't be open with your own dad."

We assume this was one of those instances when Rogers' genuine response simply bubbled up in him and he stated it out loud. While his statement may not be quite what we'd imagine one of us saying in the same situation, we can see the beauty of it. We see that he allowed himself the spontaneity to say what was in his heart, coupled with empathy for her experience. We can also see how the same statement, "You look like a pretty nice daughter," might have been meant to encourage, to try to convince her to see herself differently. Yet, we see that what makes this not the case is that Rogers' statement came from his being in touch with Gloria's inner world and being personally moved by it. If a counselor meant to convince Gloria to see herself differently by that same statement, that counselor would have erred in being more in touch with his own inner world and likely thinking how he *must* convince that client to suffer less (and hence, to experience less). This is a difference of intent that gives a statement from a counselor a deeply different meaning.

Ultimately, we want you to develop such confidence in your therapeutic listening, empathy, UPR, and judgment of when to express that you would allow true and important statements of your experience with your clients to bubble up from you, through your connection with your client's experience. The balance is to be free enough to make optimal personal connections with your clients, yet to be set and confident enough in your intentions that you don't your use significant influence to limit your clients learning from their own experiences. It is important not to take such client-limiting influence. It is also important not to let your intention to avoid undue influence cause you to think too much, to be stiff, and to lack the human connection that is key to therapeutic relationships.

A CLOSING THOUGHT ON GENUINENESS

In interviewing elementary school counselors who had been identified by their peers as outstanding in their effectiveness with children with conduct-disordered behavior, Jeff found one counselor's description of her work as representative of other outstanding counselors serving these children. She described her work as "loving them, crying with them, laughing with them, and hurting with them" (Cochran, 1996, p. 97). In his time with her, Jeff learned that strong empathy and warm acceptance were quite consistent in her relationships with such children in and out of their counseling sessions. This strong empathy and warm acceptance were clear expressions of who she was with children. When asked what advice she might give to other counselors beginning or expanding their work with such children, she first answered, "Really listen and empathize. Don't try to tell them [students with CD] what they should and should not do. Through listening and empathy, . . . lead them to find answers for themselves." She added a final thought, "Love them. If you can't love them, at least be honest enough not to fake it" (p. 106). Our desire is for you to find the strong empathy and UPR in you and to make these ways of being who you really are with your clients.

ACTIVITIES AND RESOURCES
FOR FURTHER STUDY

- Revisit the focus activity for this chapter by making a new set of observations or reviewing your notes and answers to the questions of the activity. How have your observations, understandings, thoughts, and feelings related to genuineness changed?

- Review your tapes of practice sessions, or real sessions if you are using this book while seeing clients, to search for moments that were particularly moving due to your and/or your client's genuineness. Also review for moments when you, your client, or both seemed less than fully genuine. What were the effects of those moments in the process of interactions between you and your clients, and on your clients' expressing and processing their experiences? Did your clients seem to know, at any level, when your empathy and UPR were not fully genuine? If yes, how so? Additionally, if you have the appropriate permissions, exchange tapes for review with other counselors in order to assist with these reviews of others' work and they of yours.

- Imagine moments in sessions in which you realize that your client does not perceive your empathy and UPR as genuine. Based on your learning from this and previous chapters, how do you imagine that this might happen and how might you respond once you realize it? What might you say and how? Be sure to consider reflections that go well beyond pedantic paraphrasing. What would likely be the advantages and disadvantages of the ways you might respond in such moments? Discuss your scenarios and responses with peers in small or large groups.

- Working with a group of peers or on your own, devise a tentative set of rules, explanations, and exceptions that you might use to guide when and how you would express your thoughts and emotive reactions to clients.

- Review works by Albert Ellis and others that help you understand the concept of your shoulds that may make it difficult for you to keep your empathy and UPR fully genuine with some clients. Identify your shoulds, especially those that affect how you see others. Speculate when and in what situations they may affect your work. Strive to overcome or reduce the influence of these shoulds.

- Review the literature on genuineness or congruence. We suggest you start with Tudor and Worrall (1994), Lietaer (1983), and Wyatt (2001).

- Review the Primary Skill Objectives of this chapter. Consider if you have mastered them to your satisfaction at this time. If not, reread, practice, discuss, and seek further readings until you have mastered them to your satisfaction at this time.

- Continue to review your behavior, thoughts, feelings, and communications in and out of counseling. Note when you are not as fully congruent as you would like and what seems to have inhibited your congruence in that moment. Especially in your sessions, continue to strive for fully genuine expressions of your empathy and UPR.

7

Logistics of Getting
Started with New Clients

Well begun is half done.

ARISTOTLE

PRIMARY SKILL OBJECTIVES

- Understand and be able to explain the purposes of Initial Session Reports.
- Understand the items of an Initial Session Report and how they may lead
  to treatment planning, and think through modifications you may want for
  your work setting.
- Understand and be able to explain why it is important not to let information gathering or the possible hubris of treatment planning interfere with
  your forming therapeutic relationships, and how this can happen to some
  counselors.
- Understand and be able to explain how information gathering in initial
  sessions can be necessary and helpful to ongoing work and how it could
  interfere with therapeutic relationships.
- Think through carefully and be able to explain how counseling-related
  assessment can be incorporated into therapeutic relationships and the
  difficulties that may arise in doing so.

- Understand the issue of note taking during sessions and why we recommend against it.
- Understand what should normally be included in ongoing case notes and what should not.
- Understand and be able to explain confidentiality and its limits to clients.
- Understand and be able to make an initial explanation of counseling.
- Understand and be able to explain common counselor and client problems with goals and solutions for establishing well-reasoned counseling goals.
- Be able to give examples of reasonable and unreasonable goals.

FOCUS ACTIVITY

What are your worries, considerations, and emotions around getting started with a new client? What would your worries, considerations, and emotions be in a setting in which your client asked for an appointment and you did not know until the session just what his or her concerns may be? What about for situations in which teachers, parents, or other concerned persons refer your clients to you, rather than your client self-referring? What kinds of decisions related to counseling and/or other aspects of service to this person do you anticipate making when getting started with this and other new clients?

Contemplate, journal, and discuss your thoughts and feelings in response to these questions. Anticipate setting specific considerations related to these questions for the settings where you expect to counsel. Use group discussion to learn from the expectations of peers who are or anticipate doing counseling in settings different from yours, and therefore may take referrals or handle arrangements of initial sessions differently.

INTRODUCTION

In this chapter, we focus more on the logistical aspects of getting started with new clients. We focus on structuring the therapeutic relationship, which can be a part of getting started with new clients (see Chapter 8) and then on helping clients who may need particular help getting started making use of counseling (see Chapter 9). In this chapter, we want you to think through integrating skills of information gathering, assessment, record keeping, and making basic explanations of counseling with your skills of therapeutic listening, empathy, UPR, and genuineness.

GATHERING INFORMATION
AND UNDERSTANDING FOR AN
INITIAL SESSION REPORT USING YOUR
SKILLS IN THERAPEUTIC LISTENING,
EMPATHY, UPR, AND GENUINENESS

In your initial meeting with clients, you will often have a need to gather information. You may be gathering this information for an agency requirement, for planning ongoing use of counseling, or for assessing service needs outside of counseling.

Learning about Initial Counseling Sessions from
Situations That Require Intake or Initial Session Reports

In more formal counseling settings (i.e., community agencies, large counseling centers), an intake report is often required for treatment planning. We prefer to call such reports "Initial Session Reports" because this more accurately captures how we see the first meeting with each client. The word "intake" suggests a meeting for information gathering only, information on which decisions for ongoing counseling and other services will be made. However, it is not uncommon across settings for clients to complete only one session. Thus, we would not want you to miss an opportunity to implement a therapeutic relationship that may be deeply healing while you are focused on gathering information in the only meeting you may have with a client.

In initial sessions with adolescent and adult clients, we focus ourselves in therapeutic listening, with empathy and UPR provided in a genuine manner. We usually only ask one question, something like, "Will you tell me in your own words what brings you in?" In other words, we just start counseling. Through this way of working, we find that our adult and adolescent clients almost always give us the information we need for their initial session or intake report. We find several important advantages to beginning this way with new clients versus beginning with information gathering. First, beginning a therapeutic relationship based in genuine empathy and UPR, where necessary information is gathered as a naturally occurring by-product of that relationship, allows our clients to begin their healing process from our first moments together. Second, clients are most likely to continue counseling (from this way of working), as they are already invested in the counseling relationship. Third, if that initial meeting turns out to be the only meeting we have with a client, chances are that client got or began to fulfill *her* need, rather than only giving the information *we* needed. Fourth, if our work is to be ongoing, clients are not set up with a mistaken expectation of counseling—that it will be mostly question-and-answer information gathering.

We have included the items of an Initial Session Report in Skill Support Resource C that Jeff used in university counseling center settings for you to see what information clients usually gave through the process of therapeutic listening with genuine empathy and UPR. However, if clients do not naturally offer the required information, you should stop at some point in each initial session to check whether you are getting the information you need for such a report while there is still time to shift to information gathering, if need be. When we stop to check, and certainly if we were to make such a shift, we of course tell our client what we are doing and why.

Practical Reasons for Information
Gathering in Initial Sessions

There are at least a few practical reasons for gathering information in initial sessions. A critical reason can be to assess for client safety. We will address incorporating that assessment into your work in Chapter 10. Two additional reasons include planning for ongoing counseling and planning for client service needs that you may be able to help with that fall outside of counseling sessions.

Planning for Ongoing Counseling With adolescent and adult clients, we usually get a sense in the first session whether they plan to continue or whether they expected and want only that one session (For children, we believe this is an adult decision). We've found that while most come for ongoing counseling, a small but important minority of our adult and adolescent clients come expecting only one session, often because they expect to explain their problem and get a prescribed answer. Either way, if it isn't readily apparent while listening, we strive to get a sense of our clients' expectations for ongoing counseling.

If their expectations are for a single session, we may concur that that is a reasonable use of counseling. Jeff remembers at least one client who came to tell of the sexual abuse she had suffered in childhood. She didn't want to continue counseling after that. It seemed that session had been quite helpful to her in sorting through some of her confusion and hurt. While he would have liked her to continue, and offered her this option as well as expressing his interest in continuing with her, there was no reason to question her decision for her use of counseling and to educate her for a different use. She was functioning well personally, socially, and academically. Her judgment seemed reasonably sound and her intention clear, even before starting. She came to tell what had happened to her and leave it behind, at least for the time being.

For clients whom we sense expect to have a single session in which they will explain their problem and receive a solution/advice, we first reflect our sense of their expectation. Then, we explain that we don't see that as a workable possibility in counseling. We tell such clients that we very much wish we could help them in that way, but since their and each person's situation is quite unique, we cannot prescribe a solution for them. We of course also help them understand that they can use counseling to come to understand their situation and their self better, and from that understanding reach their own more effective conclusions about how they want to handle their situation.

Through listening in initial sessions, we also try to get a sense of what each client's expectations for counseling are, meaning what is it that she wants and expects from counseling. This is an ongoing process, but if, in that first session, we can discern that a client, for example, clearly needs and wants to be listened to; has great difficulty accepting herself; is in great pain with grieving the loss of a loved one; has experienced abuse or trauma that seems so profound for her that it may take her significant time to sort through thoughts, feelings, and resulting actions; is concerned with a specific problem such as academic or social anxiety, or a specific dilemma or decision; we would like to know in order to adjust our expectations and either make explanations then or think through and offer explanations later of how this person may use counseling to assist her particular life situation. We address such explanations further in the next chapter.

Planning for Client Service Needs That You May Be Able to Help with Which Fall Outside of Therapeutic Relationships Through therapeutic listening in initial sessions (and throughout) you may think of services available outside the counseling sessions that may help address your clients' concerns and wishes. For example, if your client has a life situation for which you know there is a well-run counseling or support group, you may give him that information to consider. If you learn that your client is embarking on a job search and is uncertain how to proceed, and you know that you or others can teach job search skills, you may inform her of that. If you offer to provide this service yourself, offer to provide it in meetings that you specifically plan for that purpose. Separate it from your counseling sessions focused primarily on therapeutic listening, genuine empathy, and UPR. If you find that your clients are working to make career decisions and have come to long for information, you may arrange for yourself or others to help them find such information. If you realize that clients long for greater relationship skills and you know of workshops that are well run and may be helpful, you may give them that information to consider. Such examples of assistance that are related to counseling but not normally a part of the therapeutic relationship are unlimited, and we find that our adolescent and adult clients have often let us know in their initial session what other services might be helpful to them and whether or not they might be interested in and ready to use such services.

In a related thought, we will sometimes offer information to our clients, if we believe we know information that will be helpful to them and if they have indicated that they would want to hear the information. For example, we have sometimes given clients who seem to be deeply depressed information that we have collected regarding physical self-care (i.e., sleep, nutrition, exercise) and personal self-care (e.g., restarting previously joyful activities when possible) when they let us know that they are hurting badly enough to be ready to take action toward feeling better. It is very important not to let such information-giving veer into advice-giving or to set up an expectation that counseling will mostly be about clients receiving information from their counselor. It is also very important to watch for clients' reactions to the information and to respond to those reactions with genuine empathy and UPR. For example, very depressed clients might understandably react with dread and discouragement to information on ways to manage their painful emotions through self-care. This is no cause for counselor frustration but an emotional expression for you to respond to with genuine empathy and UPR, and an opportunity for the client to begin learning from such emotional expression and your response.

In concluding this section on gathering information and implementing your therapeutic relationship in your clients' initial session, we want you to know our firm conviction that if we only have one hour with any client, we would most want to spend it providing him with the core conditions. Even if circumstances beyond clients' control prohibit their continuing counseling with you, they will have already experienced a unique relationship with you; one that may be the spark that reawakens the flame of their drive to self-actualize. We have sometimes discovered that this new relationship is confusing to clients, and have known that to be a useful confusion. If clients believe they cannot be understood, accepted, liked, and loved but you meet them with the core conditions, even for one hour or less, that exception to their absolute belief system can be the sand that forms the pearl in their oyster. Your therapeutic relationship is a valuable gift; it is an offer of yourself in that time, and that is the best anyone can offer.

INCORPORATING COUNSELING-RELATED ASSESSMENT WITH YOUR SKILLS IN THERAPEUTIC LISTENING, EMPATHY, UPR, AND GENUINENESS

In counseling, always strive to incorporate assessment and information gathering with your skills of therapeutic listening and genuine empathy and UPR, rather than the other way around. When you realize that your task in assessing or information gathering interferes with your therapeutic listening, empathy, and UPR, be genuine and considerate and say so. As always, your therapeutic relationship with each client is most important. It is important that you establish that you are working together, that your client's desires for counseling are important to you, and that you will keep your client fully informed of what you are doing and why. If your assessment or information-gathering tasks interfere with your tasks of therapeutic listening, and providing genuine empathy and UPR, it can appear to your clients that you are hiding something or taking some expert role that they are not entitled to know about. This can appear ingenuine, can cause suspicion, and can inadvertently create a relationship where your client is unlikely to fully disclose.

THE ISSUE OF WRITING NOTES
DURING SESSIONS

We recommend that you *not* take notes during sessions. To do so would suggest that the information you gather is more important than the person you are with and the therapeutic relationship you are forming. Taking notes during a session suggests that the *information* you gather is most important because you, as expert, are going to make an assessment to prescribe an action. In contrast, your focused therapeutic listening suggests that your *client's experience* and your understanding of it are most important to you.

Jeff would like you to consider a little of his experience with the question of taking notes during sessions.

> Early in my work, I used to question my memory of session information for writing notes or reports. I have noticed that after sessions, the first thing I usually remember is how I felt with the person. Then, once that emotional memory leads me to just one detail of the actions and communications of the session, my memories of what was done and said in the session come flooding back to my awareness. Knowing this process of my memory, I have been able to stop worrying over whether I need to write anything down in order to remember it a short time later for notes and reports.

WRITING AN INITIAL SESSION REPORT

In order to give a clearer picture of how information can be gathered and may be conceptualized and written in an Initial Session Report, we illustrate examples by commenting on the kinds of things that we would likely write and how the information might come about for the items in the sample Initial Session Report of Skill Support Resource C. These items would likely need to be modified for different settings, and are meant to clarify rather than supplant any setting requirements for such a report. The items of your Initial Session Report should fit the common concerns of the clients you serve, as well as the common concerns of most people, and the information your work setting would like recorded if it is different from those concerns. Additionally, the items included on this and other well-crafted Initial Session Reports should be designed to build toward treatment planning—the last item on this Initial Session Report.

The clients for whom Initial Session Report data is suggested in these examples are completely fictitious. Rather than being based on actual persons, they are based on composites of common scenario types. In the last subsection, "Treatment Plan," they are given fictitious names, as it became too impersonal and unrealistic to write about them without using names.

Finally, we do not offer this guidance on record keeping and treatment planning because we find it essential to therapeutic relationships that heal and assist client progress. *We do not think reports are essential to therapeutic relationships.* However, they are often required forms of documentation that may serve necessary legal and agency purposes. So, we offer this guidance in order to help you get maximum therapeutic benefit out of this process when the process is required, and to assist you in not letting the information-gathering process interfere with the therapeutic relationships you form with your clients.

Identifying Information

Parts of the identifying information could be collected from an intake questionnaire or other sources (e.g., school records), but putting it in your Initial Session Report makes retrieval quicker. We usually include information that can be understood without much explanation, that doesn't take up much space or time, and that reminds us who the individual of the report is. (In situations where counselors see large numbers of clients, it can be difficult to remember each individual, especially if each person might not be seen every week.) So this would usually include age, grade, gender, and ethnicity (if known). It could include fairly permanent physical attributes, such as very long blond hair, shaved head, slight frame, or very tall. If there were some other easy reminder of the person, include that as well. For example, if a client is referred by a particular teacher or by a former client, note that (although you should not include the former client's name in the current client's report).

Presenting Problem/Concerns

This is simply a brief summary of why clients said they came to counseling. Examples might be: "upset over recent break-up with long-term boyfriend," "feelings of confusion and self-doubt related to parents' divorce," "surprised by sudden, unexplained lack of motivation for school/work, unusual tiredness, sadness, and irritability," "very strong anxiety and avoidant behavior (class/meeting absence) related to public speaking," and "lack of career direction brought to her awareness based on required sophomore career decisions."

History of the Problem/Previous Interventions

For this section, we write any history of the presenting concern that we believe may help us understand this person and concern, and inform our responses in counseling and possible suggestions of related services.

- For the client upset over a recent break-up, we would likely want to remember how long the two were together; if they have had other break-ups; if so, how this one is different; and some of the depth of meaning that this relationship seemed to have for the client (i.e., if either or both of them had pictured a permanent future together, if the client's self-esteem seem to be tied up in the relationship's continuing).

- For the client whose parents divorced, we would want to remember how long ago the divorce took place, the client's age at that time, what the parents' relationship had been like before the divorce, what it has been like since the divorce, and how the client's reaction to the divorce has evolved.

- For the client with sudden lack of motivation, we'd be curious to know how long she has had this feeling, what her motivation level was like before, and what she has tried so far to regain this motivation.

- For the client with anxiety and avoidant behavior around public speaking, we'd likely want to remember the course of development of this anxiety (i.e., has it seemed to always be there and is just now becoming a problem or does it seem quite new?), and what the client has tried toward overcoming the anxiety.

- For the client who has become aware of career indecision, we'd like to remember such things as how his career aspirations have evolved and changed, and what he has done to ready himself to decide.

In conclusion, it is the very fact that the histories are logically connected to clients' present situations that causes most clients to give such histories without the counselor asking.

Reason for Coming to Counseling Now

We have found that it often helps us understand clients and conceptualize their situations for report writing if we consider why they seem to have come to counseling that week, as opposed to a few weeks or months earlier or later. This item helps prompt us to think of what clients' expectations for time and uses of counseling may be.

- For the client upset over the recent break-up, this may be obvious—the break-up just happened, but the less obvious part may that she is worried that the break-up means that she will always be alone or is somehow a flawed person.

- For the client whose parents' divorced, this may not be so obvious if the divorce was some time ago. Upon reflection, this client or we may realize that he sees himself as currently faced with long-term life decisions and is questioning his own ability to make decisions.

- The client with sudden low motivation might be facing some deadline or may have watched someone she cares about and identifies with fail at some task that she sees as catastrophic.

- The client with anxiety related to public speaking may be facing a speech soon or may be hurting by having perceived others as laughing at him during a public speaking situation.

- The client who has become aware of career indecision may have come because of a decision deadline but may have been prompted by other events, such as a build-up of general frustration with herself and her indecision or perhaps awareness of a loved one's regret over a career decision thought to have been an error.

For a client who is facing some imminent deadline, we might expect and try a very tentative reflection of her intentions for counseling like, "I'm not sure, but I gather that since this is coming up soon, you hoped for or expected very quick work in counseling." On the other hand, for a client who seems to worry that she is flawed or is riddled with ongoing self-doubt, we might attempt a tentative reflection of her expectations for ongoing counseling with a statement like, "So far, I understand that you are really hurting with this, and that you take your situation and struggle very seriously. So, I gather you might expect our work together to be ongoing, beyond today." From either of these reflections, a more specific conversation related to ongoing work and expectations in counseling can ensue.

Alcohol/Drug Use and/or Medical Concerns

With this item, we prompt ourselves to look for potential physical complications or causes alerting us to be careful, that the client may have some physical/medical situation that would indicate counseling may need to be augmented with inpatient treatment, medical consultation, chemical dependency assessment, and so on. For example, if a client reports near overwhelming depression and heavy alcohol use, we would recognize that the heavy alcohol use may be perpetuating and intensifying the depression. Then, if that situation becomes clear and continues, we may see a need for a chemical dependency assessment.

Related Family History/Information

Family relationships are often important parts of clients' concerns, and we find that most adolescent and adult clients indicate the connections between their family history and current concerns without prompting. Also, because this connection is frequently important, we normally want to remember how each client describes this connection. Examples of the kinds of connections that might occur follow.

- The client upset over the recent break-up might explain that she sees a parent, role model, or other loved one as growing old alone and miserable, or may see a parent as character flawed or mentally ill and worry that this is inevitable for herself as well.

- The client whose parents divorced might also see one or both parents as flawed or mentally ill and assume this is his fate, or he might have little faith in permanent relationships and fear being alone.

- The client with sudden low motivation may see a parent or sibling as a failure and believe this will be her intolerable fate, or, on the other hand, she may have lived and worked hard to fill her perception of her parents' expectations, even though those expectations did not fit with her interests and abilities.

- The client with anxiety around public speaking may also believe that he can never live up to his parents' expectations or may perceive himself as already having failed in his parents' eyes.

- The client who became aware of career indecision may also have tried to live up to her parents' expectations and find it difficult to fit her interests and abilities to those expectations. She may have perceived herself as frequently criticized and unaccepted by her parents, and that may prompt her indecision.

Major Areas of Stress

When writing an Initial Session Report or, when necessary, collecting information in an initial session, considering what each individual's major areas of stress seems to be often helps us put her concerns in perspective. The client who is upset over a recent break-up, for instance, may find she can hardly think of or focus on anything else. Yet, the client whose parents divorced may be nagged by related concerns, motivated to explore them, and perhaps make changes but be even more stressed over ongoing car trouble and financial concerns.

Academic/Work Functioning

For clients who are also students, we would want to know how their concerns affect their schoolwork and vice versa. For clients who identify more as workers than students, we would be interested to know of the connection between their concerns and their work. Because persons' school experiences and occupations are often strongly related to their identities and sense of self-worth, we find that most clients offer some indication of these interactions and it is noticeable and worth reflecting on when they do not.

Social Resources

For this item, we are interested to see if each client seems to have a social support system and to what extent, and we are interested to know how each client's friendships are affected by his concerns and vice versa. Possible examples follow.

- The client who is upset over the recent break-up may have let friendships dwindle during her heavy focus on her romantic relationship or may have seemed to be isolated by a controlling romantic partner. On the other hand, she may have many friends or acquaintances who now tell her that they never liked her romantic partner anyway, that they always knew he was bad, and this may be prompting her increasing self-doubt.

- The client whose parents divorced may have relied heavily on the support of friends during the time of the divorce. He may either have strongly supportive friendships or may be experiencing great anxiety and dependence due to ongoing doubts around friendships.

- The client with sudden low motivation may seem to be using drinking and time with friends or acquaintances to avoid tasks over which she feels low motivation. This is not something we'd expect a client to say. Rather it may occur as a possibility when we notice that she has lately been drinking and spending time with friends much more. This is not a guess that we'd be likely to state in an initial session because it may be heard as a criticism, may interfere with the accepting atmosphere we want to establish, and may well be a wrongful early assumption on our part. Still, we may note it as a possibility to consider for later in the Initial Session Report.

- The client with anxiety around public speaking may also be generally shy or seem to seek out only those acquaintances that he is sure would never be critical of him.

- The client with career indecision may enjoy time with many friends who have great interest in the arts yet believe that her parents would only respect and help pay for more "practical" areas of study.

Initial Impressions or Understanding
of the Person and Concerns

In order to show you how we use this item, we offer possible entries for each of the clients about whom we have been hypothesizing through the items of the Initial Session Report.

Client Who Is Upset over Recent Break-Up Desirae is deeply hurt over a recent break-up with her boyfriend. While she is pushing herself to continue her work and other life tasks, the tasks seem difficult for her, and the break-up seems at times to be all she can think about. The break-up seems deeply wounding for her at this time for several reasons. They had been together for over a year and she had integrated him into the pictures she imagined of her future. Also she was so dedicated to making this relationship work that she has lost contact with most friends and focused the vast majority of her energy on the relationship. Now that it has ended, she assumes that it is her fault and that his no longer wanting to be with her means there is something clearly wrong with her. In spite of the great pain she feels at this time, she is continuing to take reasonable care of herself (e.g., she is experiencing some loss of appetite and craving for junk foods, but she continues to eat what seems a reasonable amount of healthy foods).

Client Whose Parents Divorced James has been dating the same person for three months, which is much longer than he has dated others. He and his girlfriend had sex for the first time recently. Since then, he finds himself attracted to and thinking of other romantic partners frequently. He explained that his father "cheated" on his mother repeatedly with other women before their divorce and realizes that his father even involved him (client) by taking him (client) with him to make it appear to his mom that that he (father) was not meeting another woman. James now worries that he will hurt his girlfriend the same way his father hurt his mother. While he has not quite said so, he seems deeply worried that there is something inherently wrong with him that is beyond his control. He explained that he has a few close and supportive friends but that he has never told anyone what his father did.

Client Who Is Experiencing Sudden Low Motivation Gina is quite surprised and perplexed by her sudden low motivation, tiredness, sadness, and irritability. She acknowledges that, corresponding with these new feelings, she has been spending more time with friends and especially going out to drink alcohol in larger amounts and more frequently. The reason for this change is unclear. However, Gina described her life up to this point as rather perfect (i.e., excellent grades, strong career expectations, ideal fiancé). Then, when I reflected that some sort of disgruntlement seemed to cross her face as she stated these things, she became tearful in stating that she was unsure whether she had ever wanted these things. It could be that she has been working toward life goals that were not her own. In an oddly related event, Gina explained that a friend of a friend died in a freak accident and she finds that she thinks about it a lot. Though unstated, through this she may have been reminded of her own mortality, and this may be prompting her to reconsider her life decisions and directions. Additionally, while she is experiencing very low motivation compared to normal for her, she is continuing to function well and does not seem to be in any work place, academic, or other danger.

Client with Anxiety Related to Public Speaking John's anxiety around public speaking has come to greatly limit his academic progress and career plans. At first, he would try to miss classes in which he had to speak publicly and accepted the lower grades. He has now dropped classes in which he must speak publicly. Yet, these are classes he must have for his major. He would consider a different major, but he already works in the business field and his employer is sending him to school to complete his degree. He explains that he is not otherwise shy, except that he will also not talk with his instructors (e.g., to explain his dilemma over public speaking or to ask for help with other matters) outside of class either. However, he explains that he is not concerned with what his instructors or classmates might think of him. He reports very high functioning in all areas except his presenting problem. He explained that he had hoped I could write him an excuse so he would not have to engage in public speaking. I explained that I cannot, and invited him to continue counseling. He reluctantly agreed to continue counseling. John appears to either have low insight and awareness, despite of his apparent intelligence, or is possibly reluctant to discuss his thoughts and feelings openly in counseling. Although he agreed to continue counseling, his reluctance and desire for some sort of permission to avoid public speaking is clear.

Client Who Became Aware of Career Indecision Chondra seems to know that she loves art and areas of study that she sees as creative. She believes her teachers give her strong indications that she has significant creative abilities. Yet, she worries that these areas of study are impractical and, while she did not clearly say so, seems to worry that her parents would not respect or be willing to pay for such areas of study. It may be that her indecision is prompted by a conflict between what she sees as practical, what she thinks her parents will support, and where she thinks her true interests and abilities may lie.

Treatment Plans

In this section of an Initial Session Report, we explain how actions within a therapeutic relationship can be expected to help the client, how we expect to respond and why, and what actions outside counseling we have or may recommend. Following from the previous examples, samples of what we might write follow.

Client Who Is Upset over Recent Break-Up Desirae's counseling will allow her to discuss the loss and discern its meanings for her. As she seems to see herself as flawed, it will likely be particularly useful for Desirae to fully explore who she is, who she wants to be, and whether she is satisfied with herself and her expectations for herself, and to make new life decisions based on her conclusions. As she discerns her views in these areas and implements her resulting decisions, she may change parts of who she is and thus fit romantically with a very different type of person from her recent boyfriend. If she concludes that the energy she committed to her friendships and romantic relationship were unbalanced and unproductive for the long term, she may decide to readjust this balance in the future. Further, the support of counseling will help Desirae to continue to value herself and maintain her self-care, and academic and workplace progress during this difficult time.

While she is experiencing very strong emotions, she also seems to be working hard not to feel. Therefore, in our work together, therapeutic listening and empathy will be very important so that she may fully learn from her current situation and grow to make the decisions she sees as best for her long-term life successes.

It will be particularly important that I communicate my unconditional positive regard for her as Desirae seems to carry great shame and sees herself as significantly flawed. Viewing herself fully and clearly may be quite scary for her. She seems to expect criticisms from others and work to hide parts of herself that she believes others are likely to criticize. I informed her of the possibilities of the Women's Support Group and the Relationship Skills Group (which is psycho-educational). She is considering these options. While there is no reason to suspect she is at any risk of harming herself or others, this should be continually monitored, since she is in significant emotional pain.

Client Whose Parents Divorced I expect James to use his time in counseling to further explore how he feels about his girlfriend, their decision to have sex, his attractions to others, the effects of his father's actions on him, who he is, who he wants to be, and aspects of himself that he sees as inherently wrong, as well as whether and which of these aspects are within his control.

It will be very important for me to listen carefully, emphasizing expression of empathy and unconditional positive regard, since James has begun to share personal information over which he is quite inhibited. Additionally, while James's emotions affect his actions, he seems largely unaware of them. Increasing this awareness will greatly inform his future decisions.

As he expressed his interest in taking actions that may help him succeed in his relationship, I recommended and he decided to pursue our Relationship Skills Group (psycho-educational) in addition to individual counseling.

While I see no reason to suspect he is at any risk of harming himself or others, I will continue to monitor for this, since he seems to have significant doubts about himself and his future.

Client Who Is Experiencing Sudden Low Motivation In Gina's counseling, her experience of therapeutic listening with empathy and unconditional positive regard will help her clarify her vague thoughts of why she is experiencing this sudden, unexpected change. I will carefully monitor for any deteriorating effects of her increased use of alcohol.

As she is now questioning whether she still wants the life goals she has worked hard for and experienced success in moving toward, this process will be key for her in resolving her doubts regarding these goals and in regaining the motivation she would like.

No need for additional resources is apparent at this time. Gina asked if I thought she should consult her physician to see if there is a medical reason for her sudden change. I encouraged her to seek this consultation, especially since she wants to pursue every possibility and reports a good relationship with her physician.

While there is no reason to suspect imminent danger or deterioration, I will continue to monitor for this, since her current situation is unusual for her.

Client with Anxiety Related to Public Speaking Therapeutic listening and empathy will be particularly important in counseling John because he seems unaware of other emotions besides the anxiety he reports around public speaking and unaware of any thoughts related to this anxiety. He may benefit from evaluating the thought patterns that put pressure on him around public speaking, once he gains some awareness of those thoughts. I may suggest that he begin to journal thoughts and feelings of the times that he is aware of feeling even mild anxiety.

As he seemed to strongly deny thoughts and feelings that I had thought seemed true for him, unconditional positive regard, acceptance of him as he sees himself, will become particularly important in order for him to risk exploring his thoughts and feelings. Additionally, he may need to spend time discussing subjects other than his anxiety related to public speaking in order to begin his use of counseling with less threatening subjects.

While John agreed to continue counseling, his commitment to it is unclear. As he lets me know him more, I will be particularly careful to help him understand how his use of counseling can help him toward his goals.

Client Who Became Aware of Career Indecision In her personal counseling, Chondra seems to need to explore her career interests and inhibitions toward decision making. As she seems easily influenced by others and doubtful of her own inclinations, it will be particularly important to convey full acceptance of all aspects of her decisions, thoughts, and feelings. Further, she seems to be trying to make decisions from a solely thought/logic approach. Therefore, empathy will be particularly important so that she may come to realize the potential emotional aspects of her decision process.

It seems that Chondra may also benefit from a greater wealth of information for making her decisions. She seems to dichotomize the world of work between "art/creative works" and "practical works." I offered to introduce and explain our resources for exploring the world of work and to sit with her while she begins this exploration. She accepted, and we set an additional meeting time for this work. I have not suggested career interest inventories at this time, as her knowledge of herself and the world of work seem immature and will develop through personal counseling, where I would expect her to discuss her reactions to what she is learning of the world of work. Further, a career interest inventory that suggests occupations or occupation types might only be an additional judgment that she accepts more than her own judgment as she is underconfident and lacks both self-knowledge and knowledge of the world of work.

Additional Notes on Thinking through Initial Impressions or Understandings of the Person and His or Her Concerns and Treatment Plans

Remember That the Process is More Important Than the Plan Never let these somewhat analytic attempts at understanding and planning for your clients encourage the hubris in which you come to see yourself as "all-knowing" or think something like, "Because I know what my client must do in counseling, I must control this person in counseling." Such assumptions would get in the way of your developing therapeutic relationships. Plus, even if you or we did know best for our clients, this would only be minimally helpful, since clients are the ones who need to know, decide, and take ownership of their actions in and out of counseling in order to maximize their unique path of self-actualization. Further, such thinking may lead you to place more faith in *your controlling* the process than you place in *the process* of providing therapeutic listening, empathy, and UPR in a genuine manner. While we often offer adult and adolescent clients some explanation or justification for the work we propose (based on their concerns), we are always accepting that our initial impressions may have been wrong. We strive to remember not to be discouraged when our therapeutic relationship with a client takes us to an unexpected solution for a problem we initially misunderstood. Rather, the resulting value to both clients and us reconfirms our faith in the power of therapeutic relationships.

Explaining the Core Conditions in Your Plan You may notice that with each treatment plan example, we of course emphasize providing a therapeutic relationship based in the core conditions. For any client we see in need of

ongoing counseling, this will be the heart of our recommendation. We wish
for you to feel free to add other aspects to the therapeutic relationships you
provide, as we have included a few possibilities in the previous examples, when
you see such additions as necessary and fitting to both who you are and who
your client is. However, we hope for you to never add something to a treatment plan that takes away from or interferes with the therapeutic relationships
that you provide. Always remember that this therapeutic relationship is the
essential core.

KEEPING ONGOING CASE NOTES

Ongoing case notes serve several purposes. They help you remember what
your client has communicated, how your client communicated it, and the
meaning or importance it seemed to have to her. They help you keep track
of how your client seems to have been doing in each session and how she has
changed over time. They help you keep track of how you responded and
what seemed to be helpful and not. They help you take time to contemplate
each client and your work with them. In the rare cases of subpoenaed case
notes, they may serve as a legal document of what you knew of your client,
what she said, and what went on in counseling. Additionally, in case you were
to suddenly be no longer available to continue or if your client suddenly
stopped attending and came back to counseling later, your notes can serve as
a guide to help another counselor know what had gone on in counseling so
far for your client.

Therefore, your case notes should include a summary of what your client
communicated to you and how, and what importance or meaning the communications seemed to have to your client. When you notice "firsts" in your
clients' actions in sessions or reports of actions outside of counseling, or other
changes, you should note this as they may demonstrate dimensions of change
(Landreth, 2002). Your notes should include comments on how your client
seemed to be doing (i.e., main affect and/or functioning). If you find that some
of your responses seemed to be particularly helpful or not helpful, you should
note this and briefly explain. Your case notes should not include assumptions
or speculations (e.g., client may have been or may be hinting at sexual abuse).
Whether right or wrong, such speculations could lead to serious misunderstandings if your case notes were used as a legal document or were needed by
another counselor assisting your client. In conclusion, it has helped us and our
students to think of keeping case notes with the following question in mind: If
you were offered a huge fortune and a blissful work situation but only if you
moved immediately and severed all workplace ties, what do you image a counselor taking over the counseling of your former client would need and want
to know? And finally, it is quite sensible to review other professionals' case
notes in your setting and guidance given by your setting regarding the setting's
preferences for case note contents.

COMMON DILEMMAS OR SITUATIONS IN GETTING STARTED WITH NEW CLIENTS

What's normal in getting started with new clients? Well, almost nothing. We find it best to review what information we have regarding a client before meeting in order to begin getting to know that person, and in case there is something we really should know first. However, we also strive to have no expectations of how each new client will be in counseling.

A Need to Know What to Expect

Having said that, we do think that most new clients have a need to know what to expect. Most clients come to counseling not knowing what opportunities await them in counseling or with mistaken assumptions. So, you may commonly offer explanations of counseling. When you offer such explanations, it is most helpful, most concrete, to begin with what your client and you will mostly do in your time together, and a very brief explanation of why. We offer examples in the next section and fuller discussions of the complexities of such explanations in the next chapter.

Anxiety

We also find it reasonable to expect anxiety in your clients in their first session. Such anxiety may be a part of why they have come.

It is also understandable to us that many clients feel anxiety about getting started in counseling. Beginning clients face the unknown of a new relationship with an unknown person. Some beginning clients react with mild frustration or lack of understanding to your actions in reflecting. This can be because, while therapeutic relationships share qualities with other relationships, counseling responses are quite unlike those of any other relationships. It may be that you are not yet artful at your tasks, that reflecting and providing other therapeutic responses do not yet come naturally to you. For you and for your client, remember that reflection and the other tasks of therapeutic relationships are *unusual* forms of communication, unlike talking to a friend or acquaintance. We have found in our work, and that of other counselors, that if counselors first accept and empathize with new clients' anxiety, frustration, or other emotional reactions to getting started and, if necessary, explain what we are doing and why, clients come to adjust and value our counseling responses.

Additionally, the physical act of coming to counseling leaves many persons anxious. Some worry over whether they have been seen and feel embarrassed. Some take use of counseling to imply that they are somehow needful, broken, or flawed.

Whatever the reason, you may expect anxiety, but also know that it may not be there. If it is there, accept it and respond with empathy. If some other emotion or way of being is there instead of anxiety, accept this and respond with empathy.

AN EXPLANATION(S) OF COUNSELING
THAT HELPS CLIENTS BEGIN

Our usual explanation of counseling, which we offer when necessary to adult and adolescent clients, goes something like, "You can think of our work together this way: Your role is to help me understand you. You may start anywhere, with whatever seems on your mind or in your heart. My role is to strive to understand you and to say what I think I understand, first probably in little understandings of what you tell me, then later with the bigger picture of how I'm learning to see you. Through this process you can come to understand yourself better and to use that enhanced understanding for new decisions and actions. Even when I am wrong in how I understand you, if you help me understand you better, you still learn."

After having worked through numerous initial sessions, some of our recent counselor interns have let us know what they have found helpful to add, at times, to our very basic explanation. Perhaps based on setting or population, a number found it important to tell their client that they (the counselor) are not judging, diagnosing, or analyzing. Another found it helpful to explain to a client that she has no expectations for what she (client) may talk about, that she (client) may think of their time together as her time, and that almost any way she may choose to use the time can be okay and will become helpful. Another helped some of his clients understand how to begin by adding his warm direction for his client to "Tell me about yourself," or "Tell me your story." A number added a quality that sounds simple but isn't always easy. They explained that in order to help clients understand how to begin, they (the counselors) needed to be as relaxed as possible with themselves. We refer you also to Chapter 8, in which we discuss more customized explanations of counseling, and to Chapter 9, in which we focus on the dilemmas that occur when clients seem to need help getting started.

Finally, while we offer examples and further discussions in Chapters 8 and 9, it is important to note that whenever possible, it is most helpful to customize your explanations of counseling to the unique person and situations of each client. Our examples of persons with various presenting problems offered under "Writing an Initial Session Report" should give you some preliminary guidance as to how you can customize explanations for each client's use of counseling. The last two items of our Initial Session Report, "Initial Impressions" or "Understanding and Treatment Plan," include how we see each client and how we expect each individual may utilize the therapeutic relationship of counseling.

INFORMATION YOUR CLIENTS SHOULD
KNOW WHEN GETTING STARTED

Confidentiality

Especially in settings where the information is not already clearly conveyed to clients, you need to routinely explain confidentiality to your clients in their first session. It is best to make this explanation brief but clear, and early, before your client might communicate anything that falls outside the limits of confidentiality.

You may need to customize your explanation of confidentiality to your client's age or other unique aspects. Your explanation must include what confidentiality is and what the limits of confidentiality are. A generic statement of confidentiality that we often give to adolescents and adults is, "Our work together is confidential, meaning I won't tell others things that you do or say here. Of course, you can talk to others about our work together if you wish. There are a few limits to the confidentiality that I can promise you. If you tell me something that gives me reason to suspect child abuse or that you are in imminent danger of hurting yourself or others, I may have to break confidentiality in order to try and get you further help or to stop the child abuse." Make such explanations flatly, meaning not mechanically or uncaringly but without implying anything more than what is said. As with any statement made to clients, watch to see how each client reacts, especially whether the client seems to understand or not, and respond to his reaction with acceptant empathy.

A customization that sometimes occurs is a need to let clients who were referred by others or seem particularly concerned with what others might know about their use of counseling that the confidentiality you promise includes the person who referred them or about whom they are concerned (assuming this is the case and you are not making an additional exception to confidentiality). For example, if a client was referred by or showed concern that a parent, teacher, employer, or spouse might have access to information about his counseling, we would let him know that confidentiality and its limits apply even to this person. In care for the feelings of the referral sources, we are careful to inform this person of confidentiality and its limits, and that it applies even to them, before taking the referral.

Who and/or Why Referred

Especially for adolescents and adults who were referred by others and seem to wonder why they have been invited to counseling, we think it best to tell them. We find that if we withhold such information when our client has indicated that she wants to know, we are being deliberately secretive and it will interfere with developing a therapeutic relationship. Again, in consideration of the referral source, this explanation is usually something we have cleared with the referring person first. We usually keep such explanations vague. The purpose of this vagueness is not to hide the truth but to avoid giving our client the expectation that we expect them to talk about a certain thing. For example, "Your mother let me know that you attempted suicide and so she and I agreed that you may be hurting and may make use of counseling." "Ms. Smith let me know that your grades have dropped quickly, and so we wondered if you are troubled and might wish to make use of counseling. In case you do, I have planned for us to meet each week for the near future." "Mr. Brown let me know that he is concerned for you. He told me that he sees you as acting out in class. I plan for us to meet weekly for the near future. I would like you to use that time to help me get to know you. In our time together you may say anything you want and do almost anything you want."

Such explanations are not always necessary. Especially, the last example seems like it might imply that this client is expected to talk about his behavior.

So, we added clarification regarding this person's use of counseling. Even if a client makes an assumption about her use of counseling based on these explanations, we find that this assumption goes away in time and that it is better than beginning with a secret about our client when our client has indicated that she wants to know. Again, it is important to make such explanations with a tone that does not imply there is more meaning to the explanations than the facts stated. Of course, it is crucial to respond to your client's reaction to the statement with acceptant empathy. If the client from the last example responds by saying forcefully, "He's out to get me and I've done nothing!" then you reflect something like, "That's the first thing you want me to understand, and you seem aggravated with the assumption that you've done wrong!" Even if this client had interrupted his counselor during the statement, we would have stopped to reflect, then finished the explanation.

Potentially Helpful Information
Related to the Presenting Problem

While it is not our norm, if we think we have information that is critical, important, and related to the difficulties an adult or adolescent client is experiencing, we may offer this information in her initial session. For example, if we believe a client is becoming so deeply depressed that the progress of the depression might overwhelm this person before she has given counseling enough of a try to even continue, we may give information that we feel confident in regarding immediate steps she may take to begin to manage the depression through self-care between that time and our next meeting. For another example, if we had thought carefully and felt sure that a particular client would benefit from assistance outside our therapeutic relationship in counseling (e.g., career information, a support group, assistance with study skills), we may go ahead and suggest this assistance in the first session.

GOALS

A final and often important part of initiating therapeutic relationships is the question of goals. Some counselors are required by settings or by third-party payers to establish goals for the counseling. In some cases, you might be required to establish goals after the first session, or soon thereafter. However, this can be problematic. Therefore, please carefully consider the thoughts we offer regarding the problems with goals in counseling and the solutions that we then describe.

Problems with Goals

On the Counselors' Part When goals are not specifically required by settings, we find that some counselors experience a need for goals that is driven more by personal-cultural habits and perhaps insecurities than from client or counseling relationship needs. Some counselors may have difficulty letting go of their comfortable role as problem solvers. Letting go of this role may feel like letting go of control. As one beginning counselor, who we know went on to do fine work, said of her first experiences counseling, "I feel like I'm in a car and they're driving." Many beginning counselors seem to believe they need goals in order to be "working" (i.e., not wasting time). Goal orientation seems predominant in our society, and we agree that goals can be quite beneficial in life. We may not be in the good places that we are if we had not frequently moved toward at least implied, if not explicit, goals for ourselves. Still, counseling is and ought to be both connected to life outside of counseling and more than what is typical of life outside of counseling. If all behavior is goal directed, which at some level it probably is, then counseling needs to be a step back from that, not just another typical life experience and relationship. Its difference or separateness from everyday life is part of what makes counseling healing when other life relationships have not been.

Goal Problems with the Client Most clients don't come to counseling knowing what goals to work on. This would require a sophisticated knowledge of counseling. Most clients just come knowing that they are hurting. Or, worse still for establishing goals, they come at the urging or direction of others versus from a more fully self-directed choice. If you ask a young adult who has just lost a romantic relationship that she expected to be lifelong what she would like to change as a result of counseling, she might well answer: "To get him back," or "To have him say, in a way that I'll believe he means and take as truth, that I am desirable and that there is nothing wrong with me, that my future will work out and I will be happy." Another hurting client might simply quizzically answer, "I want to stop hurting." Rarely would a new client answer, "What I want from counseling is to reevaluate the meanings I've made of my experiences," or "I want to evaluate who I am and who I want to be, to evaluate whether the expectations that I've placed on myself are unreasonable, and to truly examine myself, my feelings, thoughts, and actions in the light of our communication and decide what I think and feel of me, and just what I want to change and what I want to and can accept."

Further, if you tried to take a client's spoken goal and move toward it in a straight line, you may find yourself giving the young adult direction into how to get the lost partner back, or how to force that partner to answer questions that can't be answered and that could be more effectively answered from within herself, through a therapeutic relationship focused on the core conditions. But in spite of these difficulties with goals within counseling, please do not be discouraged and do read on, as there are solutions to the dilemmas.

The Thinking behind Our Solutions
to Establishing Goals

As you will remember from Chapter 1, it is our firm belief that each person is on the path of self-actualization. Further, while each person's path is unique, each person's path includes achieving the normal skills of mentally healthy persons, such as the ability to love and to work, all of the subskills within these tasks, the ability to enjoy positive social interactions, and the ability to be okay in and of her self—to be autonomous.

However, this path to self-actualization is often blocked by events and individuals' interpretations of events, or by a lack of needed nutrients for growth. This can include not having been provided the core conditions or not having had physical needs provided for, then having interpreted this lack of one's environment/caregivers as having meaning about ones' self, world, and relationship to the world.

Your role as counselor providing therapeutic relationships is to give your clients a safe and fertile environment to reevaluate attributed meanings. This reevaluation is not always or even usually overtly spoken. Your role is to provide the core conditions in therapeutic relationships, which provide a fertile ground for each individual's self-actualization growth process to restart. We have had these beliefs in the self-actualization process fostered, restarted, and enhanced through therapeutic relationships focused on providing the core conditions; and confirmed and reconfirmed over and over through our works with children and adults, and through the works of our friends, students, and coworkers.

Reasonable Goals

Therefore, we have learned that we and other counselors may set any goals for clients that are within the normal range of behavioral possibilities for the persons based on such factors as their age and abilities. Then, through counseling focused on a therapeutic relationship that provides the core conditions, progress toward these goals will almost always occur and at an optimally efficient pace for each individual.

Consider the following example of such goals:

A fourth-grade boy was referred to counseling because he had inexplicably lost motivation to do his math assignments at school and was thus beginning to fail. His parents had divorced, but he showed no overt reaction to this. In previous school years, he had made average progress and showed average motivation through his math studies. His teachers were confident that the tasks they gave him were not too difficult or too boring for him, so improved motivation for math studies could have been an obvious goal for him.

We've known numerous children in situations similar to this and could hypothesize why he improved his work through the counseling relationship he experienced. We might analyze or assess:

His experience of the divorce was one of a loss of personal power, control, or stability over his life, especially over his parents, and a loss of personal esteem, especially in his parents' eyes. While he knew how much his parents valued and worried over his math progress, he seemed to choose to fail in order to reassert himself with them and in his life. Yet, through his experience of the core conditions in counseling, he came to value himself for *being* rather than for *accomplishing* or for *personal power*. When this shift came, he became motivated to learn math again, but for the sake of his joy in learning and accomplishing instead of in order to gain approval or power.

However, while something like this hypothesis might be true, it is not something we can know or find to be particularly true, or worth spending time on. Explaining his ways of restarting that part of his self-actualization process is not as important as the fact that, like so many others in similar situations, he did apparently remove the blocks and restart his path to healthy self-actualization with the aid of a therapeutic relationship based on providing him with genuine empathy and UPR. The reasonable goal for him could have been reestablishing his interest in and motivation for learning math, with the treatment being a therapeutic relationship in which he deeply considered who he was, what he wanted for himself, and from what his value was determined.

Considering that work (we broadly define work to include anything from learning in school to being financially self-supporting or successfully taking part in the tasks of life and culture that provide for ones' physical needs) and love (we broadly define love as finding closeness and joy in relationships with others) seem so integral to surviving and thriving, we see that whenever a person is failing at tasks within these categories, improving at those tasks could be stated as goals, and we would be confident that if a goal is within the person's ability, that person will make progress toward that goal through a therapeutic relationship based in the core conditions. For the young woman who has lost the romantic partner she believed would be always with her, a reasonable goal could be to become comfortable enough in herself not to let the recent loss impede her ability to succeed at school/work and to succeed in friendships and with other significant relationships. Reasonable measures might then include improving failing grades to passing, receiving acceptable work reviews, reporting positive relationships and enjoying time with friends and significant others.

Beyond possible setting requirements, you may also want to keep reasonable goals in order to give yourself an opportunity to measure progress. This is the beginning of good and simple research. As you see clients' progress toward reasonable goals, it will help you remain encouraged in your work, which will surely be difficult at times.

Unreasonable Goals

Perhaps this is obvious, but just in case, we reiterate that goals must be reasonably within a person's ability. For example, in behavior management plans for behaviorally troubled children, we have sometimes included the goal, "Follow directions the first time told." However, this is something for the child to work toward. This is not a goal over which perfection would even be reasonable. We certainly do not always follow directions the first time told, and think that only an unthinking person would always do this.

In the example of the boy with loss of motivation for math studies, a reasonable measure of his renewed motivation would be that he seems to enjoy and completes most assignments in a timely manner. If we wanted a grade measure, it would be reasonable to base that measure on his previous grades, if his previous approach to math studies had not appeared perfectionist. However, an A grade would not be a reasonable measure if he had always been a B student who enjoyed the subject. Also, an A grade would also not be a reasonable measure if he had made A's but seemed obsessed or seemed to resent the work required all along.

The same is true for the young woman who experienced great pain at the loss of her romantic relationship. If she had seemed to enjoy work but had never been highly motivated to move up the corporate ladder, then while it might be a change she would choose, there would be no particular reason to expect that after she uses her therapeutic relationship to learn and gain all she can for herself from her loss and pain, she would then want to become a CEO.

Communicating Goals

Then, there is the question of whether or not to state goals that you record. We never like to think we are keeping something from our clients that they may want to know and may find helpful. So, if we must record goals, we think it may be best to tell adult or adolescent clients our thoughts on goals for them and hear their reactions.

What we don't like about this is that without very careful empathy, there is great room for misinterpretation of the goal suggestions (e.g., I will accept you when your grades improve), although, as always with statements to clients, we'd be ready to attend with genuine UPR and empathy to all reactions. Another thing we don't like about communicating these goal suggestions to clients is that the goals may suggest a focus for client communication in sessions (e.g., you should spend your counseling time talking of the joy you find in friendship), which is not ideal and can make for less efficient work than free-flowing client communication that is responded to with therapeutic listening, genuine empathy, and UPR. In such situations of misunderstanding, you may also need to re-explain how you see this person using counseling to accomplish these goals.

However, we conclude that if such goals are to be recorded for adult or adolescent clients, it really seems only considerate to inform them what you are recording or will record. Thus, give each client the opportunity to assert changes, if so desired by the client. Plus, we do like that communicating such goal suggestions may lead to valuable discussion of how counseling can work and how it can most effectively be used. We emphasize explanations of counseling in Chapter 8.

ACTIVITIES AND RESOURCES
FOR FURTHER STUDY

- Revisit your answers and thoughts from the Focus Activity for this chapter now that you have considered our guidance through this chapter. How have your thoughts and feelings changed? Have you any additional thoughts and feelings at this time? Do you now see some of your worries as unnecessary? Have we inadvertently suggested new worries? If your answers, thoughts, and feelings have changed, find opportunities to discuss this with peers.

- Think about times in your life that you used or can see that you might have used counseling. Complete an Initial Session Report on yourself at one or more of those times, working all the way through "Initial Impressions or Understanding of the Person and Their Concerns" and "Treatment Plans."

- Revisit the concept of self-actualization from Chapter 1 and from works by Carl Rogers and others (Bozarth, 1998; Rogers, 1961; 1980). Keep in mind your understanding of the drive to self-actualize whenever treatment planning, generating goals, or explaining potential uses of counseling.

- In a practice initial session, practice explaining confidentiality for a scenario that encompasses complicating factors introduced in the section, "Information Your Clients Should Know When Getting Started."

- In a practice initial session, practice making an initial explanation of counseling, including client and counselor roles, to a client who is uncertain and anxious over how to begin.

- Practice writing Initial Session Reports after your practice initial sessions. Discuss what you have written and thought with the persons who were in the roles of your client and observer.

- Practice generating reasonable goals based on your practice initial sessions.

- Consider and discuss with peers any adaptations of Initial Session Reports that you may want or need in your expected work settings.

- Practice writing ongoing case notes after practice of ongoing sessions. Discuss what you've written with the persons who were in the roles of your client and observer.

- After one practice of an ongoing session, try writing your notes a few days later.

- Review the Primary Skill Objectives of this chapter to see that you have mastered each to your satisfaction. If you have not yet mastered them to your satisfaction at this time, reread, seek more practice, and seek additional readings until you have mastered them to your satisfaction.

8

Initial and Ongoing Structuring of Therapeutic Relationships

> Every river has its banks, every ocean has its shores. Limitations should not always be seen as negative constraints. They are the geography of our situation, and it is only right to take advantage of this.
>
> TAOIST MEDITATION

PRIMARY SKILL OBJECTIVES

- Understand and be ready to implement the structures of therapeutic relationships in counseling.
- Anticipate potential difficulties in structuring time in counseling.
- Understand and be ready to give a general explanation of the normal structure of therapeutic interactions in counseling.
- Be ready and able to provide customized explanations of therapeutic interactions in counseling for a wide variety of persons and situations that may present in counseling.
- Discern what you expect your challenges to be in making such explanations and what you can do to succeed through these challenges.
- Consider how helpful explanations of therapeutic interactions for clients may also be helpful to significant others in clients' lives.

FOCUS ACTIVITY

Imagine yourself going to see a counselor for your first time during a time when you are upset and hurting. Assume that you know very little about counseling and that some of what you know may be unrealistic. Discuss with others or describe in writing what would be helpful for you to know about counseling. Consider what anxieties you may feel over not knowing or initially not getting what you expect from counseling. From what you know of counseling now, consider which of your anxieties it would be helpful for your counselor to attempt to assuage with structural information or explanations in order to guide your use of counseling and to help make your use of it tolerable to you, not overwhelming to the point where you quit. Imagine the information and explanations that would have been helpful to you. Also, consider which of your understandable anxieties would be best for you to work through without your counselor attempting to relieve them with information or explanation. Speculate why it might be best to work through some of your anxieties related to counseling. Then, discuss or describe several ways that you believe the experiences of others may vary from yours in regard to these questions.

INTRODUCTION

Because most clients either come to counseling not knowing what to expect or with mistaken expectations, one of your tasks as counselor is to structure or clarify clients' use of counseling. This includes basic logistics as well as explaining what each persons' best use of counseling may be.

LOGISTICS

Session Length and Ending Sessions

Some of the logistics of adolescent and adult use of counseling that we find important include the length and frequency of sessions, and issues related to scheduling and rescheduling. Regarding the length of sessions, we usually inform our adult and adolescent clients, "Our meetings will normally begin on the hour and end 10 minutes before the next hour (This would vary with settings, of course). I will let you know when we have about 5 minutes left in our meeting." As with all statements we make to clients, we watch to see how our clients react and then respond with empathy. Normally after this statement, clients look at their counselor as if to say, "Okay, that's fine." To such a look we respond, "Seems like that makes sense and is okay with you."

On hearing this advice for letting clients know the time structure of counseling, one of our beginning counseling students exclaimed amusedly,

"Oh, I wondered why my counselor always ended before my hour was up. I thought he was cheating me out of 10 of my minutes."

Time Warnings and the Importance of Letting Clients Own Their Endings

Within this explanation of session length is the structural information that you will give a five-minute time warning. These time warnings and generally ending your sessions on time allow your clients ultimate responsibility over use of their time in counseling. For example, if your client is working up to telling you something before the end of the time, it is helpful for your client to know when his meeting with you is nearly ended for that day. Or, if your client has been quite emotional in the session, he may choose to take time to dry tears and pull himself away from the strong emotions before returning to class, your waiting room, or other less private places.

To maintain your client's right to choose how she uses her time with you, it is also important to end your sessions on time. We have sometimes found that clients want to tell their counselors about parts of their lives over which they may feel shame or embarrassment very near the end of a session in order to say it with no time remaining to see or hear their counselor's reaction. We think it is true for many of us humans that no matter how much we've come to trust others, we'd sometimes rather not see their reactions when we tell them the things of which we feel most ashamed.

An extreme example of this comes from early in Jeff's work.

I had worked with a young man for 10–12 sessions. I am confident that he had perceived the core conditions from me. He had shared with me parts of his life that were important to him and difficult for him tell me. In the time that we worked together, he was recovering from self-cutting and other self-injurious behaviors. Just before we began working together, he had left his family's church, to which he had belonged all his life. He helped me understand that he had internalized many self-criticisms from the teachings of that church and from his perceptions of his parents' criticisms of him. His parents had broken off contact with him after he left the church, but he and they were working to reestablish a relationship over the weeks that he worked with me in counseling. His girlfriend, whom he had very strong feelings for and had hoped to marry, had broken up with him shortly before he left the church. His life had been in turmoil. During our time together, he had worked hard reviewing and reconsidering many of his former self-critical beliefs and self-doubts. When we began, he'd seemed quite fragile. By the time we ended, he was strong and seemed to be getting stronger each day. Much of his process had been to tell me things from his past over which he felt shame or guilt. It seemed to me, as he told me of these actions and worked through his self-criticisms, doubts, and feelings of shame and guilt, that he had come to deeply trust my respect for him and for his judgment, and that he was taking clear responsibility for and ownership of his use of counseling. I thought he had finished this work, as his time of telling me such things had seemed to reach a crescendo, then ended. He had also known well in advance when our ending would be.

Then in the last five minutes of our last meeting (we had known that it was definitely our last session, as we would be unavailable to each other after that time), it seemed to me there was something more on his mind. I reflected this. He paused and hesitated. He told me again how much he appreciated our time, how much he'd grown, and how much of a support our work had been. I knew that he meant these things and that there was something more he was ready to add. Then in our last moment, he told me that he had pressured his girlfriend to have an abortion a year before and that I was the first person he had told. He said this at the door and left immediately, leaving me no chance to respond. I took this to be a huge disclosure, as I knew that his former church would have considered it a great sin.

But it became clear to me that this was his choice of how to end, that it was no accident. For myself, I wanted him to stay. I wanted him to have more time to discuss his reactions to the abortion. I wanted to hear more of his reactions to the abortion. Yet, my role had been to respect and accept him as he presented himself to me. This included his decisions of how to use his time in counseling. I thought quickly as to whether I should seek him out for safety concerns, but I knew that he was future oriented, strong and gaining in clear-headed strength, and had given me believable, solid reasons why he had stopped any self-injurious behavior and would certainly never suicide (we had discussed these things earlier in his work). If it had been possible for me to call him back, this may have conveyed to him how much I cared. Yet, I was already confident that he knew I cared deeply. So, if I'd had an opportunity to call him back, it would have communicated that I did not trust him and his decision of how to end, from which he may have generalized that I hadn't trusted others of his decisions. Then, because he had made it clear to me that my views were very important to him, he may have concluded that his self-doubts and self-criticisms had been deserved after all.

That experience serves as a strong reminder to me to remember to allow clients to know how much time we have, then to stick to my commitment of that time. Endings of sessions and counseling relationships are important. They belong to your clients. As I was able to rule out imminent danger concerns, I needed to trust his decision of how to end, of what he wanted to tell me in our last moment and how. Such trust had been my role with him, and it would have been a violation to change in his final moment, even as I wanted to know more and do more for him. As you structure your counseling consistently, your clients are empowered to use their time, including endings, as they see best.

Exceptions to Ending on Time

As with any guidance, there are always exceptions. An example of an exception to ending on time comes from a client that Jeff first mentioned in Chapter 2.

> In an initial session, she'd explained that she had something big to say and that she just had to get it out. She told a story of sexual abuse in her childhood and explained that she'd never told anyone and it was hard for her to say. At 45 minutes into her session, instead of giving a 5-minute warning, I interjected something like the following, "We are nearing the end of our normal meeting time length. It seems that you are in the middle of telling me of a part of your life that is very important to you right now. I can extend our time a half-hour if you wish." (I was fortunate not to have another meeting scheduled immediately after hers. If I had, I'd have had to schedule a time to continue.) She agreed and stated her appreciation. Near the end of our extended meeting, I gave time-structuring statements as usual.

Another exception to ending at 50 minutes, and the only obvious situation for which counselors should allow themselves to be made late for a following meeting, would be sessions in which risks of imminent danger are unable to be resolved by the normal ending time. When such situations occur, you have a legal and ethical obligation to resolve safety issues before your client leaves. While this is true, we still don't like the message this may give to the client for whom you are late—that the time of persons who are in crisis may be more important to you than those who are not. So, avoid this exception to ending on time whenever possible by addressing issues of imminent danger within the normal meeting time (Managing situations of imminent danger is addressed in Chapter 10).

The Awkwardness of Giving Time Warnings and Guidance for This

Some of our beginning counseling students tell us they feel awkward learning to give time warnings. They worry that the time warnings stop their clients' communication and perhaps make it appear that they (the counselors) are more interested in time than in their clients' communications. We offer a few thoughts for such situations.

- A five-minute time warning can be paired with a reflection. An example is, "We have five minutes left today. You were telling me that while you feel uneasy, you are proud of the break you have made."
- Whether a five-minute warning is paired with a reflection or not, sometimes it does stop a client's communication. Sometimes clients will change the subject to discuss rescheduling or ongoing work. This is not necessarily a reason to assume that this meant they thought you were disinterested. Perhaps they simply saw a need to take care of planning for ongoing work at that time.

- Once your clients are used to your routine, you can usually give time warnings quietly, without even stopping their communication to you.
- Endings have to come sometime. Endings based on time are artificial. Yet, our culture and its institutions are usually based on coordinated time use. Thus, for you to make your time available to a number of persons, you must use the artificial endings based on time.

Possibilities of Varying the Time
Warning Structure for Some Clients

Again there are always exceptions. For example, you may choose to give a client who has difficulty ending first a 10-minute and then a 5-minute warning, in order to give that person more time to prepare to end.

A Few More Suggestions on This Time Thing

Place Your Clock Conveniently Place a clock you can easily read in your field of vision so you can be aware of the time without having to be highly distracted or having to make a distracting motion to check the time.

Realize Your Keeping Time Maximizes Your Client's Freedom to Express You are keeping track of time so your clients do not have to. Rather than being restrictive, your keeping track of time is freeing to your clients. Because you keep track of and give time warnings, your clients are free to focus their attention on their self-expression and use of counseling.

Losing Track of Time There will be those sessions where you simply lose track of time. This is usually due to having gotten so caught up in your client's experience that you simply forget about the time. In those cases, once you realize your mistake, simply do the best you can to approximate a compromise that allows for time warnings and ending close to on time.

HELPING CLIENTS UNDERSTAND THE STRUCTURE OF INTERACTION IN COUNSELING OR HOW COUNSELING MAY WORK FOR THEM

The primary and most frequent structure of interactions between counselor and client is this: A client tells the counselor about his self and life; the counselor reflects the part of that communication that she saw as most important to her client; the client reacts to reflection in continued communication of his self and life. Some clients come expecting such interactions. In a great many cases, clients learn this structure of interactions and its value simply through participating in it, without explanation.

You may remember our base explanation of counseling that goes something like, "Your role is to help me understand you. You can start with any part of yourself and your life. My role is to strive to understand and to say back to you the parts that I think I understand, first little pieces, then a larger picture of who you are." We sometimes add this explanation of one purpose of this process: "You can then use my understanding of you to increase your understanding of yourself. When I am wrong, you may come to a more accurate understanding of yourself by correcting my misunderstanding. When I am right, it may spur deeper self-understanding on your part. You can then use these new understandings for improved decisions based on knowing who you are, and who and how you want to be."

This explanation is imperfect, not entirely accurate. We use the word *understanding,* when what we really mean is *empathy.* This is because empathy may be harder for clients to understand. Yet, "understanding" is usually readily understood. Also, this explanation does not capture the essence of therapeutic relationships, which is each client's perception of genuine empathy and unconditional positive regard from his counselor. While these qualities are more important than the description of new insight in the added explanation of purpose, the explanation does reasonably describe counselor and client interactions, and gives clients a useful initial idea of what counseling can be like.

A very helpful way to improve such explanations, especially when clients continue to struggle with how to use counseling, is to customize your explanations to individual client situations. Examples follow.

A Client Who Asks for Guidance

We have served a great many clients who have become accustomed to being controlled by others yet, especially as an adolescent or young adult, resent this control. The following is an example from Jeff's work.

> I was working with a young man who was referred by his physician for stress-related health concerns. He had described his father to me as very controlling and communicated that he felt great pressure to perform for his family. He was studying in a very difficult field, which he explained he did not have much interest in but knew that it was what his father wanted and his family expected.
>
> I felt his intense frustration with this dilemma. At first, when he in exasperation asked for my guidance regarding how to solve the dilemma, I reflected, "That's how *frustrating* it is for you, so much that you really want an answer from me." When, after further expressing his frustration, he confirmed that he really would like guidance from me, I responded by briefly reflecting, "That really is what you want." Then continued explaining, "I'd like you to think of counseling working this way . . . [I offered a version of our explanation of roles in counseling, from the previous section]." When I saw a look of dissatisfaction cross his face, I reflected, "I get the idea that you were bothered by the suggestion for

counseling that I just gave." He responded, appearing to make great efforts to maintain respect for me while expressing his dissatisfaction, "Yes, I need help. I don't know what to do."

At that time I gave further explanation, customized to his situation and our therapeutic relationship. I explained with empathy, "I'd like you to consider . . . [I paused, then continued speaking slowly and carefully because his perception of my words were very important to me in that moment] First, I don't believe I can know what is best for you to do. I believe that each of our situations is quite unique. But even if I did, then I'd worry that you'd be trading taking direction from me for taking direction from your father, whose control you want to move beyond." A look of recognition and surprise crossed his face, which I knew meant that what I'd said was new but beginning to make sense. I responded to that recognition and surprise and continued a part of my explanation that I believed also important: "It seemed like my last idea made sense to you and sparked a new thought. I also want to add that I have great faith in the counseling process and strong confidence in *your* struggling through and making *your* decisions."

He responded that he really didn't want my suggestions, that it was just how difficult the dilemma was for him, and that he really did want to decide for himself. We worked together a few sessions more. He decided to openly oppose his father's control and family expectations. He quit college to take time to decide what he would like to do before returning. To do this, he had to give up financial support from his family.

If I had given into my temptation to give him advice when he asked, I would have probably led him to pick a major that he liked more and to talk to his father about this. However, his decision was that a more abrupt, clean break was necessary for him. Once he made that decision, it became obvious to me that it was best for him. I came to see that it gave him the best chance of keeping his resolve now and then reconciling with his father later, having built a strong sense of self-respect. My advice, based on my worries of how he would suffer without his father's support, might only have prolonged his agony. Fortunately, I disciplined myself to trust him and the counseling process, to meet him with empathy and unconditional positive regard—the most powerful and useful tools that any counselor can have. Oh, also his health problems had also dissipated before our work together ended.

A Client Who Seems to Insist on a Quick Solution

An example of a client who also asked for guidance but who seemed entrenched in insisting on using counseling only for guidance occurred from a client who was served by an intern in Jeff's practicum class. The young man was referred from a health service that saw his symptoms as depression (e.g., sleeping excessive amounts, continually feeling tired, and disinterested in activities). He had begun with one counselor, insisting that he must have immediate relief. When that counselor responded by trying to convince him that he should consider medications, he was angered and asked for another counselor. He had tried medications in the past for depression, and his father took medications for depression but seemed to only have ongoing problems.

When he was switched to his second counselor, an intern in Jeff's class, he began by explaining that he was willing to admit he was depressed but was adamant against considering medication. Rather, he wanted to know what he could do to make the bad feelings go away besides using medication. He seemed almost panicked for a quick solution. It may have been that while he was not feeling hopeless yet, he feared that if he did not find a solution soon or if medications did not help, it would mean there was something uniquely wrong with him and he would then be hopeless. He seemed quite resistant to exploring himself and his emotions. He seemed determined to focus on searching for a quick solution.

In class, we helped his counselor prepare to provide him with information on self-care for managing emotions (i.e., managing sleep, diet, exercise, and use of enjoyable activities such as hobbies—such information can be found in a large variety of sources, such as Bourne, 2000). However, we knew that while he was asking for such skills, they alone would be much less powerful than the work he could do with a therapeutic relationship in counseling. So, our supervision group generated a variety of explanations of his possible use of counseling. In this way his counselor could offer one or two from among them when the time was right in his sessions. We include several of them here for you to consider the variety of helpful forms such explanations can take.

One student counselor noted that the client's work with his current counselor would be very different from what it seemed like he had experienced before in that there would be no judgment or criticism of his decisions from his counselor. His counselor noted to our supervision group that the client had repeatedly said "how dumb" his thoughts were. We offered the explanation that in counseling, without judgment or criticism, he could come to see his ideas and himself for who he really is, not who he or others think he should be. In that way, he can get to know parts of himself that he really likes and prizes, and get to know some parts that he would choose to change. Then, as he comes closer to being a person that he generally likes and prizes, by emphasizing the parts he already likes and changing others, he may find that his painful feelings lessen.

Another intern suggested that it might be important to add that in his (the client's) process of gaining greater self-understanding through helping his counselor understand him, he may have to face emotions that are quite hurtful and hard to express. We thought this addition both a true expectation and helpful as it let him see his work as significant.

Another intern suggested reflecting the client's urge to make his painful emotions go away, then adding a thought from his counselor's perspective that an opposite way of being and an option in counseling is to accept and acknowledge emotions. We added the important thought that, in this case, while his counselor would not be solving his emotional pain for him, he also would not be facing it alone but would have his counselor right there with him.

It occurred to us that this explanation might understandably lead the client to ask why he would want to accept and acknowledge emotions that hurt and that he wants to be rid of. To answer such a question, we decided that his counselor could explain that emotions are useful. Like physical pain, there is something to be learned from them. For example, if a toddler touches a hot burner, the toddler learns that it hurts and to avoid that action again. While physical and emotional pain are rarely that simple, there is something important about one's way of being to learn from it. Therefore, emotional pain is not something to make go away but is an important tool for finding needed guidance.

We also thought that the client seemed to be working hard to solve his painful emotions, and we thought of offering the metaphor of trying to untie a difficult knot. Sometimes, the more a person works at it, the tighter it becomes. Then, if the person relaxes with the knot and explores it without trying to untie it, either parts of solutions simply come or you find that the knot has somehow already loosened in your hands. By this, we mean that if he uses counseling to explore his whole self, without trying to make his painful emotions go away, his solutions to his painful feelings may become obvious or the feelings may begin to dissipate without clear solutions ever being found.

We also noted that he seemed to have had a couple of recent experiences of going to experts for help, then disagreeing with or even being hurt by their suggestions. Thus, he could use counseling in which he explored himself, his feelings, and his ideas as a way to renew valuing himself.

The client opened his next session with a renewed request for guidance on how to make his painful emotions go away. His counselor first reflected that guidance for managing emotions was what he saw that he most needed and then offered ideas on ways to manage difficult emotions. The young man seemed disappointed and expressed his doubt and frustration that these changes sounded very difficult and didn't sound like they would be strong enough to help him feel all that much better. His counselor carefully reflected his emotions and conclusion within this statement, paused, then added that she also wanted him to consider using counseling very differently than he had so far. She gave one or two of the explanations of counseling for this client that we had offered in class. The young man seemed to like the empowerment he felt with this way of looking at his use of counseling. He of course continued to struggle, as neither his life nor use of counseling was simple (we don't believe anyone's is). However, he did make a strong use of counseling and found strength and self-confidence through facing his fears. Through this process, he came to feel less depressed and his symptoms of depression decreased. He chose and made changes in his life that were satisfying to him.

A Client Who Has Great Discomfort with Silence

There are silent moments in counseling. As discussed in Chapter 2, this is not necessarily a problem and may be a good thing. Sometimes you will reflect, your client will pause not knowing what to say next and feel awkward in that silence. Jeff once served a young woman for which this was dramatically true.

The customized explanation that helped her may serve as a useful example to consider in customizing explanations for others.

Some background may help in understanding the customized explanation that follows. In our (the client's and my) first couple of sessions together, she had discovered and let me know that she very much feared being alone, so much so that she would sleep with men to avoid being alone. She had come to know that these men didn't care for her as a person and would usually not be there for her after sex. After one awkward silence in a growing pattern of such moments, I reflected, "You seemed to come to a moment when you had nothing to say and feel awkward with that." She responded, "Yeah, it feels really weird." She looked at me with an expression that seemed pained and meek. I reflected what I felt from her, "I gather that even beyond weird, the awkwardness seems to hurt and you want it to go away, maybe for me to do something about it." She continued with moist eyes, "Yeah, I can't stand it. Is this the way counseling is supposed to be?" I reflected and explained, "Sometimes it seems you just can't stand it. [I paused, then continued] Yes, there are sometimes very awkward silences in counseling. While I know they are very difficult for you, I think you may help yourself greatly by working through them, by finding the thoughts and feelings that come in those silences."

This explanation seemed enough for her for a while. Later, when a silence had left her feeling particularly awkward, she again looked at me with a pained expression, near tears, and exclaimed in a low voice that seemed to also convey that she felt defeated, "I can't stand this. I'm not getting anywhere." I reflected and then explained, "It seems it's starting to seem too much for you." I paused and saw the agreement in her eyes, then added, "I think that even as hard as it is, it may be that some of the same thoughts and feelings that you know will hurt so much when you are alone may come up in these silent moments. So, I think for you, counseling may be an opportunity to face those thoughts and feelings, as if alone but not alone, in the safety of our counseling time."

In response to this, she seemed to gird herself up to continue. She moved on to tell me of a time that she had tried being alone over the past few days, some of the critical thoughts of herself, her family, her future, and the pain that had come with those thoughts.

A Client Who Attends Sporadically

While we accept clients choosing to end their work in counseling, either permanently or just for a while, we also know the value of attending counseling consistently. Jeff's work with the young woman in the previous section and his structuring explanation regarding her sporadic attendance may serve as a useful example.

She fell into a pattern of attending sessions for a few weeks in a row, then missing from one to a few weeks. When she was in counseling, she would seem to make progress. When she would come back after missing, she would return with renewed hurt from her behavior and the actions of others. At the beginning of one returning session, I offered a high-level reflection of her behavior and my reaction, which also served to help her understand how she was using counseling and its effects, saying, "I'm glad to see you again. I worry that you come to counseling for a while, begin to face some very hard thoughts and feelings, then leave and stay away for a time." I paused to see her shrug what seemed like tentative agreement. I continued, "Then, when you come back, it seems that you have gotten hurt more in the time you were away." She tearfully agreed this was true and told me of her most recent and devastating emotional hurt. She continued more consistently over a longer period after this interchange.

I wish I could say that my explanations to her had produced a great and simple change in her. In truth, I'm not sure. While she made great progress in counseling, it seems that other forces and events in her life proved potent in comparison with counseling and related resources. In the time that I knew her, I continued to worry that her story did not seem headed for a happy ending.

A Client Who Just Does Not Know Where to Start

Some clients simply do not know where to start or what to say in counseling. This can be because the client was talked into trying counseling by someone else (e.g., a parent, a spouse or romantic partner, a roommate, an employer, a teacher, a physician). In this case, your client may come to counseling perfectly willing to use counseling but still have great difficulty getting started. We have noticed that often this is based on the client thinking she has to be a good client, which she has interpreted as consistently talking about a problem and something that seems deep. Our work with children in child-centered play therapy (Axline, 1947; Guerney, 1983; Guerney, 2001) has helped us know that actions, or in this case words, that seem like nonsense can spiral into great existential questions (i.e., Who am I? Who do I want to be? What is my value? Can I measure it—if so, how?) within the process of a therapeutic relationship. Therefore, we have sometimes used explanations like the following:

Client: [Pauses after a reflection, looking at her counselor with questioning gestures.]

Counselor: You seem to have run out of things to say and are looking to me to help you out.

Client: Yeah. I'm sorry. I want to work on this, but I just don't have anything . . . [hesitating pause] that was all I could think of.

Counselor: So, you're working to think of something to say but coming up with nothing.

Client: [Laughs a little.] Well, I think of things to say, of course, but just stupid stuff, not for counseling.

Counselor: Oh, so it sounds like ideas cross your mind, but you rule them out for counseling. I'd like you to know first that a pause in our conversations may not be a bad thing. Sometimes out of awkward pauses, the most useful things occur. Also, you don't need to worry whether the things that occur to you to say are worthy of counseling. You can say the first things that come to mind and the process of our interactions will take us to the topics that are important for you.

There are countless more varieties of customized explanations of individuals' use of counseling. For just one more example, Jeff has often helped clients who want to become more personally expressive or self-reliant see counseling as a safe place to experiment with being different. In summary, the following are the keys to helping clients who seem to need explanation to understand how they can use counseling:

- To meet each client with empathy and unconditional positive regard, as much as possible, of course
- To not take a client's struggle with counseling as an indictment of you or your counseling
- To come to an initial understanding of who each client is, what his situation is, and what he seems to want. Then offer an explanation, based on that understanding, of how he may use his therapeutic relationship with you to help himself.

HIGHER-LEVEL REFLECTIONS CAN ALSO HELP CLIENTS UNDERSTAND HOW TO USE COUNSELING

Higher-level reflections, in this case reflecting your clients' process or pattern of behaviors in counseling, can help your clients become aware of how they are using counseling and often make unnecessary more extensive explanations of how they can best use counseling. The added awareness of how they are using counseling gained from your process reflection helps your clients to make optimal choices for their use of counseling.

Example 1

If you have a client who consistently waits until near the end of her sessions to get to what it seemed she wanted to say—the topics that have the most emotional content and seem important to her—you can reflect this to her. An example of how you could say this is, "I notice that in many of our meetings you seem to start by telling me about your week and only get to topics that are emotional for you shortly before our time is up." Just as you accept what your client feels, it is important that you be ready to accept her process in counseling. It is important that your tone in making process reflections remain accepting. The rationale for acceptance, as always, is to empower your client's self-responsibility and healthy decision process. However, in this situation, acceptance has a situation-specific importance as well. Consider that the client in this example may greatly fear getting to the things that she came to communicate. She may be building to communicate still more. If so, it is important for you to honor the pace at which she is ready to self-disclose. Pushing her to go faster than she is ready and able will only destroy trust and inhibit progress. She might get to what she wants to say sooner but only with your controlling her, which would be a hollow, half victory at best. However, reflecting how you see her using her time allows her to be aware of the pace she is choosing and alter this pace when ready and able.

Most clients simply acknowledge the reflection and continue their communication while adjusting their process based on their new awareness. Some clients also question or comment on your process comment. For example, clients might respond to the aforementioned process reflection in a variety of ways. They might ask, "Why, is that not okay?" To which you would reflect and explain, "You're not sure if it's okay, and you want it to be. It can be okay. I want you to work at the pace you are able and with the topics you see as most important. But I also know that the sooner you get to what you most want to say, the better."

Another client in this same scenario might also tell you, "I know I'm slow. It's just so hard for me to say emotional things." To which you may reflect, "You see yourself as slow and want me to know that that's because it's very hard for you to tell me things that are emotional for you."

Still another client in a situation like this told her counselor, "Yeah, I try to give you all the background you need to understand, then it seems like we're almost out of time." In this case, her counselor reflected and explained, "Oh, so you're thinking that I need the background to understand the important part. I'm thinking that may not be necessary. You could try just starting with the important part and I may be able to understand without as much background."

Example 2

Another example of a situation for which a process reflection can also function to help clients be aware of how they are using counseling is a client who tends to express emotions and communicate regarding a topic that seems clearly important to him, then quickly changes the subject to talk about something less emotional and seemingly less important to him. In this case, your process reflection could be, "I notice that you tend to begin to tell me about situations and feelings that seem very important to you and to feel strong emotions, then quickly change the subject to something less emotional."

Again, many clients will simply acknowledge the reflection and continue their communication while adjusting their process based on their new awareness. And, some clients will comment on or question such a process reflection. We've known clients to say, "Yeah, that's what I do. I try not to think about hurtful things too much [sounding light and satisfied in tone]." To which you can respond, "So you know that's a pattern for you and you seem satisfied with it." Ironically, since this client did seem to convey satisfaction with his statement, hearing the acceptance in your reflection, he might say, "Well, not satisfied—its just the best I can do. It seems to work for me . . . but the painful emotions just seem to come back. . . ." In this case, your reflection has allowed him to question his pattern with the full awareness prompted by your attention and empathy and within the safety of your acceptance.

To that same process comment, another client might respond, "Yes, this is very hard and I can only deal with so much at a time." To which you might respond, "So this is hard and you see that you are working at the pace that you are ready and able." Again, accepting this client's judgment and communication is the quickest way to allow her to continue to trust you and to intensify her pace. Also, it respects her decision to maintain her slower pace if she continues to decide that that pace is optimal for her in that it is not overwhelming or so painful that she would quit trying.

EXPLAINING THERAPEUTIC RELATIONSHIPS OR USE OF COUNSELING TO SIGNIFICANT PERSONS IN CLIENTS' LIVES

It is often helpful to explain counseling to significant others in clients' lives. Versions of the same explanations you might give clients are helpful here, too. We give examples of elements of explanations of counseling for parents here, then address other aspects of consultation with significant others in clients' lives in Chapter 13.

If a parent refers or brings her child to counseling, we often think that the parent has a right to know what kinds of things will go on in her child's counseling or at least may need to know what kinds of things go on in order to be supportive of the counseling. Elements of explanations that we often find helpful in such situations follow:

- "Your son may use counseling to come to better understand himself, then use that new understanding for new and improved decisions about how he really wants to be and act in life."

- "Counseling can be quite difficult and painful. For any person to fully understand himself, he may have to face parts of himself over which he is highly critical or ashamed. This can be true whether or not he actually has a rational reason for self-criticism or shame. So, your son may not always be eager to come."

- "If your son has acted out, the misbehaviors may have been an attempt to hide these parts of himself. If your son has acted in self-destructive ways, it may well be for the same reasons."

- "As your son comes to see himself clearly, he will come to act differently based on accepting who he is or changing parts of himself in order to become the person that he really wants to be."

- "I ask for your support in being ready to listen with an open mind and heart when he chooses to talk with you about himself or his counseling."
- "I also ask you to support him in not asking for information about his work in counseling, until he chooses to offer it."

Just as explanations of counseling to clients should be customized to their unique situation, so should explanations of counseling to significant others in clients' lives.

PROBLEMS FOR BEGINNING COUNSELORS IN EXPLAINING CLIENTS' POTENTIAL USE OF COUNSELING

Our interns have told us that they know the kinds of explanations of counseling that we suggest are true. They know this from their experience as clients in their own counseling. Still, during their first sessions as counselors, sometimes the explanations do not come to them. Usually, once their early experiences of success lead them to be more relaxed and calm as counselors, this problem goes away.

Sometimes, not thinking of helpful explanations that you know are true comes from getting too caught up in your client's drive to find an immediate solution to feelings that hurt or in your client's despair over not knowing what to say, perhaps not trusting her own judgment in self-direction. It helps in such situations to take a step back and try to see the whole person in context, rather than just her immediate demand or plea for a solution, or perhaps her immediate frustration with counseling. For example, consider again the young man who was pleading for help with his depressed feelings but was also working hard to make his depressed feelings go away, with no success. His counselor was having difficulty thinking of how to help him understand how he could successfully use counseling, because all the counselor could hear was his plea to make the painful emotions go away. When his counselor was able to take a step back, she saw that the young man was not only demanding to make the painful emotions go away but was doing and communicating other things as well. He was working to avoid feeling, he was discounting his own views and asking for those of others, and in working so hard to untie his knot, he was actually pulling it tighter. The overall metaphor of this paragraph is the old expression, "can't see the forest for the trees." In this case, the counselor was staring so closely at a single distressed tree—the young man's demand to tell him how to make the painful emotions go away quickly—that she was unaware of other facets of her client's communication that would be helpful to express.

Spend careful and focused time with the Activities and Resources for Further Study for this chapter to work through problems that are likely for you in making helpful explanations of counseling.

ACTIVITIES AND RESOURCES
FOR FURTHER STUDY

- In a 15-minute practice session(s) in which your client (a classmate or peer) communicates to you on a topic that has emotional content for her or him, add the skills of time structuring to your growing repertoire of counseling skills. Alternate roles. When in the role of client, after the session ends give your counselor feedback, including what her or his way of time structuring left you feeling and thinking as client.

- Imagine difficult clients for yourself, both some who are similar to examples in the chapter and others who may be different. Have partner(s) role-play those clients while you work helpful explanations of counseling into their sessions. These practice sessions will probably need to be 20 minutes long in order to give the opportunities for explanations time to come up and to be reacted to. In this case, your partner will probably have already given you feedback in the reaction that she or he acted out. However, you will still be helped by stopping to hear further feedback after the session, as opportunities for such feedback from actual clients will be rare.

- Imagine scenarios where you explain counseling to significant others in your clients' lives. Have partner(s) role-play significant others in practice meetings in which you work in helpful explanations of counseling. Again, your partner will probably have already given you feedback in the reaction that she or he acted out. However, you will still be helped by stopping to hear further feedback after the session, as opportunities for such feedback from the significant others in your clients lives will be rare.

- Write about these activities in your journal. What was difficult for you? What came easily? What do you expect will be difficult for you with clients and significant others in their lives? What will you do to work through these difficulties? What do you expect will be your strengths in such situations? How will you capitalize on your strengths?

- Discuss again or rewrite your response to the focus activity. Notice how your thoughts, feelings, and statements on the topic have changed and/or stayed the same.

- Review writings by authors, such as Virginia Axline (1947), Louise Guerney (1976, 1983, 1995), and Gary Landreth (2002), for counselor behaviors they describe in child-centered play therapy and other works with children that may be considered structuring. Consider how children come to understand, use, and take control of the structures they are provided through play therapy. Then, consider how this may help you think differently about structuring when counseling adolescents and adults.

- Revisit the Primary Skill Objectives for this chapter. If you have not yet mastered them to your satisfaction at this time, please reread, engage in additional practice, and seek additional readings and study opportunities until you have mastered them to your satisfaction.

9

When Clients Need Help Getting Started

It is easier to resist at the beginning than at the end.

LEONARDO DA VINCI

PRIMARY SKILL OBJECTIVES

- Understand and be able to explain several reasons why some clients may struggle in starting.
- Understand and be ready to avoid each mistake that a counselor might make that could inhibit a client's beginning use of counseling.
- Understand and be able to implement each of the examples of counselor actions that may help clients struggling in starting.
- Understand and be able to explain the roles of empathy and UPR in helping clients begin to fully utilize therapeutic relationships with you.
- Understand how and what types of questions or topic suggestions might help clients who are struggling and what types would not. Be ready to implement this understanding.
- Understand and be able to differentiate and explain the parts of your clients' struggles in starting that are within your influence, the parts that are not, and how you may best respond to each.

FOCUS ACTIVITIES

Activity 1

Think of a variety of clients with situations that might leave the persons struggling to start to make use of their therapeutic relationship and perhaps needing some help getting started. Strive to understand the reasons why some clients might struggle in starting. Examples might include persons who are generally shy and rarely talk about themselves, persons who feel very anxious but have not realized what they are anxious about, persons who usually spend their time helping and focused on others, or persons who were urged into counseling by others and do perhaps feel badly but have no idea why or how to start. Most clients do not need help getting started, but for those who do, there are an infinite variety of reasons why. Try to put yourself in the situations of each of the persons that you imagine might need help getting started. Try to feel what you imagine each person might feel and think, and act as each might think and act.

Contemplate how you might naturally feel, think, act, and be with clients who are struggling in starting. Then, imagine things you might do or say, as their counselor, or ways you might be with each person that you expect would be most helpful in their struggle. Journal and find opportunities to discuss your contemplations on these matters with your peers.

Activity 2

Imagine times in your life when it would have been difficult for you to get started in the counseling processes described in previous chapters. Why might it have been difficult for you? What parts of your self or situation might have gotten in the way? What might you have felt and thought, and how might you have been and acted? What might your counselor have done to help you through that time, if anything? Journal and discuss with peers your thoughts, feelings, and answers to these questions regarding such a time in your life.

INTRODUCTION

In this chapter, we focus on ways you can help those clients who struggle in beginning to make use of their therapeutic relationship with you. We find that when we and other counselors follow the skills of the core conditions of counseling as we have described them through the preceding chapters, most clients readily and rapidly begin their work in counseling: expressing and discovering their self, reviewing and redeciding their ways of being. Yet, some struggle in starting for a variety of reasons. Some may be inhibited by negative preconceptions of counseling. These clients may have internalized a perception, stated or implied by family, friends, or society, that using counseling means there is something terribly wrong with them or that they are weak and not self-reliant. Or they may have internalized the perception that counseling is disempowering, that the counselor will break them down with analysis, expose their weaknesses, and assume a long-term need for ongoing therapy. Others may fear that they have something to hide and that their counselor will see and critically judge this part of their self. Some may be shy and not at all used to relating closely to another person. Some may have anxieties that nag at them and that have driven them to counseling but that they avoid feeling fully and facing at any length. We humans are naturally taught to avoid anxiety because it is, at the very least, uncomfortable. Yet, in counseling, it will be hard to avoid facing painful anxiety. Some clients may have often been judged, have felt hurt by this, and greatly fear critical judgment from you, as their counselor—a person who will come to know them deeply. The reasons that some clients may struggle in starting seem infinite. We would like you to consider that while it is a minority of clients who struggle in starting, their struggle can be quite understandable.

When such slow-to-start situations occur, we urge you to first review your ways of being for common mistakes that may be inhibiting your clients' progress. Then, after considering the possible effects of your ways of being, consider statements and explanations that may help struggling clients understand how they may use counseling, and lastly, consider the limited use of questions that may help struggling clients get a stronger start.

MISTAKES YOU MAY MAKE THAT INHIBIT YOUR CLIENTS' BEGINNING USE OF COUNSELING

Trying Too Hard or Worrying about Motivating Clients to Make Rapid Use of Counseling

Jeff's grandfather once added to the old expression, "You can lead a horse to water, but you can't make him drink," the additional phrase, "but a goat will eat anything." You see, with goats, the problem is not getting them to eat or drink but containing them and limiting them to eating and drinking the things that are healthy for them. Part of what he meant was that some people focus too much on the problems of motivating others and come to see those they wish to motivate as a horse that simply will not drink. Papaw was a good motivator of people, and one of his ways of doing this was to recognize when he could see them as goats, not horses; when he could and should give them some very limited direction, then get out of their way, having provided an environment that is fertile and safe for their work. A therapeutic relationship is an optimally fertile, safe, and healthy environment for your clients' growth. So, while there may be things you can do to motivate clients to start their rapid use of this relationship, the first and most essential thing you must do to help your clients who are slow to start is to *accept their pace as it is* and *trust the process* of providing a therapeutic relationship based on genuine empathy and full UPR.

Not Recognizing That Your Client Began
to Use His or Her Therapeutic Relationship with
You As Soon As You Began to Provide It

It is very important to accept your client's pace and actions in counseling. If your client is nervous beginning counseling, does not talk as much as you might expect, and has uncomfortable pauses in her early work with you, then rather than indicating that she is not using her therapeutic relationship with you, it means that working through those uncomfortable pauses is an important part of how she *is* using her therapeutic relationship with you. You might offer such a client help getting started in the form of explaining how counseling may work for her or even offering questions to suggest topics. However, these guiding actions may not even be necessary when you provide genuine empathy and acceptance for where your client is and how she *is* starting.

An Example with a Quiet Client Imagine that your client seems quite uncomfortable with repeated pauses, and you feel just how hard and scary getting started is for him. In that case you might reflect your client's process by saying something like, "You seem unsure of what to say and this leaves you very uncomfortable, apprehensive, maybe even afraid of how this will go." Because his experience is happening within the healing environment you are providing, working through this discomfort and fear in getting started is how he has already started and may be a very important part of his use of the therapeutic relationship you provide. It may be his working through this anxiety in beginning to express himself with you, while your empathy helps him experience his anxiety fully and your acceptance helps him to be okay with experiencing, which builds his confidence to face other challenges, helps him see the things he fears of himself in a more realistic perspective, or leads him to experience his anxiety fully and learn whatever he needs to learn from it.

An Example with a Talkative Client Another client who seems slow to start may begin, not by talking less but by talking more, while seeming to express little that is important to her. Some clients who are anxious about starting their therapeutic relationship with you may try to fill all the time with words that seem to express anything but their heartfelt experience. Persons in this situation might talk about others in their life, their to-do list, or the weather. Again, this is not really a problem. After reflecting core content and any attached emotions, if it is coming to be how you see a client, you might tentatively offer a process reflection like, "I get the idea that there are things you'd thought you might say or do here, but you seem to talk of anything but those things." If necessary, you could even add a structuring explanation for this client's use of counseling, such as "While I want you to use this time to discuss any of the things on your mind and heart, I think it may be most efficient if you consider what thoughts and feelings are most important to you and express those. While I don't want you to focus on editing what you say, I also want to help you get to the core things that will be most helpful for you to say." However, in most cases, nothing needs to be done to help such clients, except to focus yourself in therapeutic listening, especially emphasizing empathy, which will lead to the heart of the matter, and accept this as her start. Just the process reflection and following with therapeutic listening, deep empathy, and genuine UPR will usually suffice. Example dialog of how this might go, following the same process reflection from before, may clarify our guidance for such situations:

> **Counselor:** I get the idea that there are things you'd thought you might say or do here, but you seem to talk of anything but that.
>
> **Client:** [Responds with a slight giddiness that seems to really be a mild nervousness, and with an apologetic tone.] Oh gosh, I didn't mean to waste your time, it's just, I thought I was supposed to say anything.
>
> **Counselor:** Oh, you seem to have taken my statement like I was fussing at you. Yes, absolutely, you may say anything. [Counselor combined a quick reflection with a structuring explanation that seemed important to avoid misunderstanding.]
>
> **Client:** [Sounding slightly exasperated and picking up pace again.] It's just, I'm so busy and have so much on my mind that I wouldn't want to waste time here. I'm sorry.
>
> **Counselor:** Oh, so you were talking quickly through subjects in order to not waste a minute. I also get the idea that you worry slightly that I am critical of you. I see that you are working hard. [Counselor combined three brief, related reflections that each seemed important.]
>
> **Client:** [Sounding more emphatic] I am working hard and no one seems to appreciate it. I am sick of it! [This was her strongest statement yet.]
>
> **Counselor:** [Responding with same emphatic tone] You've about had enough of people not valuing your hard work!

The conversation may then go to a trend in wanting appreciation and seeming to get criticism, or perhaps to lists and explanations of hard works, but our purpose in offering this quick example is to show how staying with such seemingly slow-to-start clients can lead to strong starts and beginning expressions that are close to your client's heart. In this example, the client did not respond to the parts of the counselor's reflections that were about their relating in that moment. So, we would assume that those parts were too much too soon, somehow scary, or just not the most pressing things on her mind to express in those moments and so were not key to her immediate experience in that moment. In fact, it may never be necessary for her to have direct conversations about her relationship with her counselor in order to make full use of that therapeutic relationship.

Lack of Acceptance

Sometimes counselors can care so much for their clients that their way of being communicates, "I will accept you when you show me that you are using our therapeutic relationship" versus "I accept you now and know that you have already started using your therapeutic relationship with me, as you are."

The behaviors of that mistaken way of being often include having a single picture in mind of what it will look like when clients are making efficient use of their counseling relationship with you, then responding enthusiastically to client behaviors that fit this picture and responding with feelings of anxiety and disappointment to client behaviors that do not fit that picture. This beginner counselor error is sometimes exacerbated by counselors' insecurity and belief that either their clients must be making clear progress or it means that they are bad counselors, or by counselors seeing their clients as fragile and thus not acceptable as they presently are.

Pedantic Reflections

Another reason your clients may start slowly is that you probably have not yet become artful at therapeutic listening and providing a therapeutic relationship based on the core conditions. This is a set of tasks that sounds so very simple but that we and many of our friends and peers expect to devote lifetimes to perfecting.

For example, in the previous short dialogue example, the counselor sometimes forgoes reflecting content, which can be overly pedantic and repetitive, to reflect what the counselor thinks he sees the client doing; mixes in a brief structuring statement; maintains strong empathy; and seems to know what is important to emphasize. Developing such a skill level will take devoted practice time and experience. Yet, everyone must start somewhere. Every doctor has her first patients and every counselor his first clients. While it may be easier said than done, we suggest you accept yourself as the counselor you are now. If that means that your work is not as efficient as it soon will be and that some of your clients get off to a slower start, we are still confident that you and your clients will progress more quickly if you accept and work where you are, instead of overly criticizing your work and trying to rush ahead to be the counselor that you are not yet.

Reflections That Sound Like "Aha" Conclusions

Slipping into trying to be a more artful counselor than you are so far can lead to giving too much weight and exuberance to your reflections when you think you've figured something out. For example, you might say, "So you feel *disappointed* when you *think* you're unappreciated because you *thought* you deserved it." Now, there is nothing wrong with and a lot right about such a response from a counselor. But if it comes out sounding like you just thought of the key word that will help you complete the crossword puzzle or like you are *naming* what your client is going through, rather than *experiencing* it with him, your client will be stopped in his tracks from experiencing and expressing. This mistake usually comes from trying to rush ahead to get to something important (as if every action and expression in counseling were not already important) or "from coming on big when you're feeling small," trying to be the expert you think you must be to be respected, when that kind of respect is really not important.

Slipping into Questions

Along the same lines, if you feel anxious with a client who seems slow to start, which is probably normal, you may slip into asking questions that give your client the impression that you mean to lead her through an investigation of herself or her problem that leads to you spitting out sage wisdom for her. You may have begun to use questions with this client, just to help her get started, but with both of you feeling anxious, it got out of hand. When you make this mistake, your client will probably stop expressing freely and wait for your next question or sage advice. This client is putty in your hands and thus dependent on you for molding. Unfortunately, putty rarely gets to live an active, autonomous life.

When in Doubt, Please Review

Whenever you wonder if there is something you are doing that may be slowing your clients' starts, please review Chapters 1–8 for other aspects of skills that you may enhance in order to better facilitate your clients' starts. We have organized them to make them ready references for you.

EXAMPLES OF COUNSELOR ACTIONS THAT HELP CLIENTS STRUGGLING IN STARTING

We developed the following suggestions with a class of beginning counselors who had just succeeded with an unusually high percentage of clients who seemed slow to start in counseling. Many of the clients were willing to be in counseling and had things to work on like all of us humans but came to counseling at the urging of others, instead of with a clear internal motivation. A few may have been hostile to making use of counseling, due to defensiveness or inhibition. Each of these suggestions was a solution in helping unique clients work through struggles in starting. We offer them for you to consider, as you carefully develop your own solutions for the unique situations you face with clients. Also, we don't want to imply that the purpose of any of these suggestions was to ameliorate the struggles of clients. The purpose is to strike a balance between allowing each client to gain as much as possible from his struggle in starting, while helping him with his struggles so that he is able to tolerate counseling long enough to work through and learn from his struggles.

Start Where Your Client Is

Acceptance is always key. One client was willing to try counseling for personal growth but may have been unsure where to start. At times, she tended to talk about topics that her counselor understandably saw as superficial. Her counselor explained in supervision that he was at first put off by this, but then realized that as he trusted the counseling process and especially accepted his client where she was, the seemingly superficial issues grew to more personal and personally meaningful topics. For example, a client may begin discussing being devastated by a broken fingernail, then that topic may grow into discovering that the broken nail was a small event but one that prompted her built-up stress and emotion to overflow, and she may discover the value she places on appearance, or a tendency toward perfectionism, or a low tolerance for unexpected happenings.

We think it very important to add to this suggestion that this counselor could not accept his client's description of her devastation over a broken nail in order to have her move on to topics that might be seen as deeper. This would not really be acceptance or starting where she is. Rather, accepting any client with her feelings at a given time or over a given event is valuable and effective in and of itself. *The healing aspect of acceptance is acceptance.* The point is not to help your clients move to topics that you see as valuable but to experience and accept your clients' values and feelings in that moment. Through your acceptance of their feelings and thoughts in that moment, your clients can come to better accept themselves and their values, change if change is needed, and grow in the directions that they choose in the light of their full consideration, brought by your genuine empathy and acceptance of them just as they are. Your acceptance and your clients' self-acceptance over broken fingernails or any other event that you or some counselors may see as small may very well be the key to self-improvement for many of your clients.

Respond to the Level of Emotion Each Client Expresses

Beginning counselors often think of clients as having started when they express clear or strong emotions, but small expressions of emotions are sometimes missed. So, it is helpful to clients who are struggling in starting counseling for the counselor to respond to even low levels of affect present in clients' communication. Some emotions are always present in human communications. For example, writing this section is mostly a thought process, yet we also feel *hopeful* that you will find it helpful and feel a *mild concern* for you as you face challenges such as clients who struggle in starting.

Reflections of small emotions that may help you see what we mean follow. If a client is telling you about a housemates' dirty dishes but really doesn't seem troubled by the situation, you may respond with something like, "So while it's on your mind, you're really okay with that." To a client who is talking about his sister more than himself (when *you* might wish him to focus on himself), you may respond in a tone that conveys both the acceptance of his communication that you are striving for and your empathy for the mild emotion he is expressing, worded something like, "As you tell me about her, I get the idea that you have a mild worry and dislike for how she is doing those things." To a client who is discussing future plans but only expressing mild emotion regarding them, you might respond, "So, you have some concern and some excitement regarding these possibilities." In each of these situations, the client would probably not have overtly stated the emotions, but the emotions would have become evident to you through the way the communications were made and through the emotions you feel in empathy. This, of course, once again brings up our suggestion that you continue to uncover and redevelop your natural abilities for empathy.

Small emotions lead to stronger ones, if strong emotions need to be expressed. Then, focusing on your clients' emotions helps them to express the core of what they wish to express and to express the core of themselves to you.

Remember the Uniqueness of Each Client's Pace

There is no parallel path that all clients must take in counseling. One client may seem to start slowly and another to start very quickly. But then, the client who seemed slow in getting to "important" topics may more quickly apply what she is learning from counseling. Some clients seem to gain insight into themselves from counseling, then readily apply what they have learned. Some clients seem to quickly gain insight and then struggle to apply what they have learned. For some clients, it seems to take a long time to start becoming self-aware, but once the self-awareness starts, it grows like a snowball rolling downhill. Other clients never seem to have any profound insight or at least never state it in a way their counselor recognizes yet quickly make behavioral or emotional changes outside of counseling. For efficiency, your goal should not be to help them change but to meet them with the therapeutic relationships you provide.

One beginning counselor explained in supervision that he realized in the moment he was telling a client that he had no expectations for what she might say that he really did have expectations for her. That client came to struggle less in starting as her counselor faced his expectations and came to accept her unique pace.

Remember That It's Natural to
Feel Uncomfortable in the Beginning

Uncomfortable feelings, often even anxiety, are common in the beginning for both counselors and clients. New counselors may be doing the work for the first time and understandably may feel at least some concern over whether they will do well and be perceived as doing well. New clients may very understandably fear possibly being judged, criticized, analyzed, or taken to emotional places they do not want to go. Both are meeting a new person and, especially for the client, are in a new setting, and both are starting a new relationship full of unknowns.

So our advice, which is simpler to say than to do, is if this discomfort or anxiety is present, accept it. The more you struggle against it, for yourself or for your client, the more it will grow and the longer it will last. The sooner you accept it, the sooner you may reap its benefits and move on to other challenges.

Give Them Room

We like that one beginning counselor phrased it as giving clients who struggle in starting room to figure out what they need and how to use counseling. By this, she meant a similar thing as starting where your client is but also implied toning down her own excited anticipation of profound change. We would add that in counseling focused on providing therapeutic listening, genuine empathy, and UPR, there is nothing else to do but self-discover and change in ever-maturing directions. So stay tuned in, but sit back and give your clients and the process room to work.

Respect Your Client's Pace

Along with giving your clients room and starting where each client is, it is important to not push and to respect each client's pace. It may be an irony that the more you push, the slower your clients move. At one time, Jeff worked with quite a few women clients who had been raped or sexually abused by men.

> I realized that I was being particularly careful to respect their pace in counseling and their decisions. I was thinking that they had already been very hurtfully disempowered by men, and I wanted to be particularly careful that each of these clients felt safe with me as their counselor and a man. Later, when I realized their rapid use of counseling, the lesson to me was that there was no reason not to respect their pace and were many empowering reasons to maintain this same respect with all my clients.

See the Big Picture in Your Clients' Communication

When your client seems to go from one topic to another and does not seem to communicate any one with depth, it can be most helpful if you look for the big picture in her communication and reflect themes you see within her communication, such as commonalities between topics (e.g., irritations with others, a concern over how others see her, a focus on the future or past) or the emotions with which the topics are mentioned (e.g., "These seem to be all things you are okay with," "I notice you seem to feel worry over a variety of situations," "You seem to carry hurts from the past and to be cautiously excited about the future," "These thoughts of the future begin with excitement, then seem to carry an edge of apprehension"). In seeing the big picture, you may also reflect your clients' process (e.g., "I notice you tend to begin to talk about one subject or another, then stop," "I have the thought, when you change topics, that there is something you are not saying or that you think to say but don't want to right now," "I notice that when you seem to begin to feel uncomfortable emotion, you end with 'but that's okay' and change the subject"). In seeing the big picture, you may notice incongruencies in your clients' presentations or mismatches in things they tell you (e.g., "I remember you had said that you prefer not to dwell on old hurtful situations, but I see that they also stay on your mind," "I know you told me you had some hard decisions to make, and I notice you haven't talked about those much," "I'm not sure, but I get the idea that you are hurting over some things, and it seems these may be things you haven't told me yet").

It is very important when reflecting the big picture you think you see in your clients' ways of being with you that you say what you see in an accepting way. Otherwise, such statements come across as sounding like a criticism (i.e., perhaps to a client who is already sensitive to criticism) or like an analysis. Your purpose is to let your clients see what you see in them, through your lens of empathy and UPR, and then to make choices informed by that view of them. Such reflections may prompt questions from your clients regarding how they are supposed to use counseling. You would, of course, first reflect what your client has communicated to you with her question, then, if necessary, answer how you see her and what you might suggest for her use of counseling (see Chapter 8 for examples of explaining clients' potential use of counseling).

Dispel Expectations of a Need For a Problem or a Profundity

Some clients try too hard to be good clients. So, part of an explanation you may offer to some clients who struggle in beginning counseling is that their use of counseling does not necessarily mean there is something wrong with them or some big problem that they need to work on. Also, if they seem to be struggling to find something profound or worthy of your and their counseling time, it may be helpful to let them know that they may start with the first things that come to mind, that they need not edit what they might say in order to find something worthy or seemingly important, that speaking their mind and heart without editing can be the most efficient and helpful way they can use their time, and that when they communicate without editing, the counseling process will take them where they most need to go.

Sharing Experience in Letting Go

One beginning counselor decided to tell her client who struggled in starting that she had recently made use of counseling for herself and found it an "experience in letting go." This seemed to help her client in a number of ways. It allowed the two of them to share a possibly similar struggle. It normalized her client's struggle. And it introduced the concepts of letting go of control, editing, and forethought in her use of counseling.

While this worked quite well for this client, we caution you to use such self-disclosure sparingly, if at all. We find it can turn the topic to the counselor, away from the client, and it assumes a sameness of experience that is unlikely given the uniqueness of each person and our diverse situations.

And Finally, Hang in There

One beginning counselor explained that she had worked to accept her client as she was and to start where she was, had prompted with questions and possible topics, and had explained her client's possible use of counseling different ways. But this slow start seemed to last forever. Then, in the third session, her client "just started" and seemed to work rapidly. Then a little later, this

counselor realized her client had started from the moment they first interacted, but it was in the third session that their discomfort dropped and it
became clear her client was working.

ASKING QUESTIONS OR SUGGESTING TOPICS THAT CLIENTS MAY FIND HELPFUL IN THEIR STRUGGLES IN STARTING

Normally, we prefer not to use questions or suggest topics. This allows clients
their most self-directed use of counseling and avoids creating misunderstandings of the therapeutic relationship in counseling (e.g., that it is a relationship
in which the counselor keeps the power and responsibility by choosing the
topics and asking the questions that eventually lead to a counselor-driven solution or suggestions). Yet, we are willing to compromise with this preference in
order for adolescent and adult clients (we do not find this necessary with children) to work through their struggle in starting and to not quit counseling.
So, we provide guidance here on offering questions or suggesting topics to
help clients who struggle in starting.

In extreme situations, after reflecting a client's great struggle in starting, we
will offer to suggest topics in order to make getting started bearable to a client.
It may go something like this:

> **Counselor:** [First reflects.] You seem to be struggling to find something
> to say and you seem to feel really uncomfortable.
>
> **Client:** [Shrugs with helpless, pained look.]
>
> **Counselor:** [Reflecting again] It seems this struggle is downright painful
> for you.
>
> **Client:** [Nods acknowledgement and shrugs again with helpless, pained
> look.]
>
> **Counselor:** Look, while I want you to work through this struggle, I also
> don't want the struggle to be so hard that you just hate coming to
> counseling.
>
> **Client:** [Gives another smaller nod of acknowledgement and that helpless,
> pained shrug again, as counselor speaks.]
>
> **Counselor:** [Continuing] So, to help you get started, to find the things
> you might like to talk about, I can try asking you some questions or
> suggesting things to talk about.
>
> **Client:** [Relieved, but still struggling] Oh, okay. Thanks.
>
> **Counselor:** So, you would like that [then offers a question or topic
> suggestion].

Stating Why A Client Was Referred
or Why You Offer Counseling

Sometimes an obvious topic to suggest for clients struggling in starting counseling is the reason why they were referred or why you offer counseling. So picking up at the end of the previous vignette, the counselor might offer, "I know that Ms. Smith is concerned for you, and so she suggested to me that I offer you this time. I believe she cares for you and has thought you seemed troubled lately." Or in another situation, the counselor might offer, "Your mom has the idea that her divorce from your dad, or something about their situation, has been hurtful to you." After either of these example statements, we as counselor would wait to see how our client responds and reflect that before making any more specific suggestion. More specific suggestions will likely be unnecessary, at least in that moment.

Jeff remembers a situation with a child, "Mike," in which it seemed dramatically obvious that he needed to just tell the boy why he was hoping to spend time and offering himself to listen.

> I knew that the father of this boy had killed himself. When Mike and I sat down, we both were silent for a moment, then I told him, "Mike, I know your father died [Actually the details of the death were not clear to me, although suicide seemed certain. Still, I stuck with saying only what I knew]." I didn't state any assumptions of what Mike might need or feel. I just stated the little information that I knew and stopped to wait. He then spoke of how he found his father's body, and he became understandably emotional. I was later glad that I'd not offered any assumptions of how he felt, as I learned that, of course, his relationship with his father had been quite complex, and he felt a confusion of emotions, even more than the usual, powerful emotions of grief.

Suggesting Common Areas
of Importance for Discussion

There are topic areas over which most people feel strong emotions and have a complexity of things we might express. These topic areas may include: school/work, family or family of origin, romantic relationships, friendships, current stressors, and future plans. You should only bring up one of these at a time and select the topic area you think might be most useful to your client. Then, phrase your suggestion with only limited explanations and assumptions. Following from the vignette offered near the beginning of this section, the counselor might say, "Well, as I know you are a student, it seems you might tell me some of your thoughts and feelings related to school." Or the counselor might suggest, "I know that romantic relationships are sometimes important to persons about your age [knowing that the adolescent or young adult is single], I wonder if there are any thoughts and feelings related to that that you might find useful to talk about?"

Basing Questions/Suggestions on Information
That You Already Have and Are Interested In

If working in an agency or counseling center, you will probably have written information about your client before you start, such as a checklist of common concerns that people bring to counseling. In such situations, we have taken this written information as if it were our clients' first communications to us and so reflected the parts that seemed most important to our clients to help them begin face-to-face communication, when necessary.

In schools, you will also usually already have some information about your clients before starting counseling. So for example, you may be able to offer a topic like, "I know that your grades have dropped this year. I wonder what thoughts and feelings you have related to that?" Or "I am aware that you have changed schools a number of times. I imagine that has meant making new friends and adjusting to new people a number of times. I wonder what it has been like for you?" Or if I knew that a client had used counseling before and had already given me this information in writing or a statement, I might be curious and ask, "I know that you did meet with a counselor before. I'd be interested to hear about that."

Open Questions

If you use occasional questions, keep them as open as possible. By this we mean that the possible answers are as varied as possible. If we ask, "Was meeting with a counselor a useful experience for you before?" this is a very closed question, as it suggests a yes or no answer. An in-between example would be, "I know that your grades have dropped. I wonder what it is that might be troubling you." This question carries an assumption that will likely shape the client's answer—that something is troubling him and this causes lower grades. Another modification of a closed question is multiple-choice: "I wonder if your grades may have dropped because something is troubling you, or do you find your subjects harder this year, or do you just like them less?" We find this type question as hard to follow as multiple-choice tests. Closed questions limit clients' attempts to communicate to you. These last two closed question examples also suggest that you are moving into a problem-solving mode, which will also limit your clients' taking responsibility in their communication to you and time with you.

CONCLUDING THOUGHTS
ON HELPING CLIENTS WHO
ARE STRUGGLING IN STARTING

Consider these closing words of advice for the times that you serve clients who struggle in getting started.

- If anxiety is there for you, your client, or both, accept it and experience it. It is an important part of your therapeutic relationship in that moment.

- Know that while you may make compromises and help a client who struggles in starting, your responsibility is not to *get* your client started but to be with your client as that person struggles and as struggling *is* that clients' way of starting.

- Whenever possible, maintain a safe, confidential place and time in supervision or another relationship to give yourself an opportunity to voice your frustrations and search your ways of being for elements of your being with clients that inhibit your clients' use of counseling. It has been very helpful to us and other counselors we know well to be able to acknowledge or admit when we are having trouble accepting a client. Realizing our lack of acceptance through safe, confidential supervisory or peer relationships has been a great help in growing our acceptance.

ACTIVITIES AND RESOURCES
FOR FURTHER STUDY

- Revisit your answers to the focus activities and consider how you might change or add to your answers at this time.

- Role-play scenarios in which clients struggle to start. When in the role of counselor, pay particular attention to how you feel, what you do to work through your feelings, and what you do to help your clients with their struggle in the moments that you find it necessary to compromise their responsibility and attempt to help your clients with their struggle. When in client roles, pay attention to how the situation feels for you, how you feel when your counselor seems to experience your struggle with you but does not offer assistance, and how it feels when your counselor does offer assistance [Note that negative client feelings do not indicate poor counselor responses any more than positive ones. Remember that sometimes clients need to struggle in order to take responsibility]. When observing, carefully consider the effects of the counselors' actions. After each role play, discuss the decisions the counselor seemed to make at various points and the likely long-term outcomes of those decisions.

- Consider the common counselor mistakes that inhibit clients from beginning to use counseling. Which do you think will most likely to occur for you and why?

- Consider what might be your greatest difficulty in serving clients who struggle in beginning their therapeutic relationship with you.

- Consider what might most prompt anxiety in you around clients struggling in starting, and discern how you will work through your anxiety.

- Design a set of advice for yourself for such situations. Word it in ways most meaningful to you and focused on the skills you anticipate needing in those situations.
- Review the Primary Skill Objectives for this chapter. If you have not yet mastered them to your satisfaction at this time, please reread, engage in additional practice, and seek additional readings and study opportunities until you have mastered them to your satisfaction.

10

Managing Client Crises with Therapeutic Relationship Skills

In the middle of the journey of our life I came to myself
within a dark wood where the straight way was lost.

DANTE ALIGHIERI

What did I know ... What did I know of love's austere and lonely offices?

ROBERT HAYDEN

PRIMARY SKILL OBJECTIVES

- Understand and be able to explain how therapeutic relationships create power and influence to help clients manage crises.

- Understand and be able to explain the roles of empathy, UPR, and genuineness in helping clients manage situations of imminent danger.

- Understand and be able to explain the meaning and importance of each principle for managing client crises with therapeutic relationship skills.

- Understand and be able to explain the meaning of the assessment factors in determining level of risk for a client who seems to have thoughts of suicide.

- Understand and be able to explain non-self-harm agreements, including what they are, what they are based on, how and why they can work, and how they are influenced and enhanced by therapeutic relationships.

- Understand and be able to explain pros and cons of hospitalization for mental health safety and how any difficulties can be eased.

- Understand and be able to explain assessment factors and aspects of your therapeutic relationships that can be used to help clients determine their level of risk in situations in which the danger is domestic or dating violence.

- Understand the common difficulties for beginning counselors in helping clients manage situations that may be of imminent danger.

- Develop reasonable confidence for helping clients manage situations of imminent danger with your therapeutic relationship skills, through reading, discussion, contemplation, and practice activities.

- Be reasonably ready to help clients manage situations of imminent danger with your therapeutic relationship skills, through reading, discussion, contemplation, and practice activities.

FOCUS ACTIVITY

Part 1

Try to remember a time in your life when you were at some risk of intentionally (intentional at any level of awareness) hurting yourself or others, and/or a time when you were at some ongoing risk of being hurt by others, at whatever age these situations may have occurred for you. It is not unusual for persons to have had thoughts of suicide at some time in life. If you have had such thoughts at a time in your life, remember what it was like for you at that time. While it may be more rare than serious thoughts of suicide, many people have at least had strong thoughts of hurting others cross their minds. If you have had such thoughts, remember what it was like for you at that time. Often through dating, domestic, or family violence or potentially violent situations, many persons have experienced being in at least some danger of being seriously physically hurt by others in their lives. If you have experienced such risks, remember what that was like for you.

If you can recall such risks being a part of your life, remember what your thoughts were at these times, not just the thoughts that one might understand as potentially dangerous but the other thoughts that you remember crossing your mind near the time too. Recall what you felt at the time, both physically and emotionally. Recall how you acted, including your actions that did or did not seem related to the risk but that you now see as possibly related. Recall how you were, meaning your qualities of being, such as going blankly or numbly through the motions of daily tasks; keeping overbusy and perhaps avoiding thinking and feeling; seeming weighted, like moving through molasses; or perhaps seeming highly sensitized and hyperalert.

After stopping to fully consider what it was like for you at these difficult times, list and explain the things that you think a counselor or caring other did or might have done to help you. Also list and explain things you believe such persons tried or might have tried that were not or would not have been helpful. In order to more fully understand your experience and to begin to further your understanding of what such situations might be like for others, journal or essay, then if possible, discuss and share your experiences with others. Take some time to let others know what it was like for you and to hear what it was like for them. It makes sense to us that you or your peers might feel inhibited or fear each other's judgments of the situations you share with one another. This is not uncommon, even though ideally we think such fears and inhibitions should be unnecessary. So, please participate in these discussions as much as you possibly can while also maintaining the privacy that you want or need at this time.

Part 2

Whether or not you can remember such risk-filled times in your life, consider what you imagine such times may be like for others. What do you expect different persons might think and feel, and how do you expect different persons might act and be at times when they have thoughts of hurting themselves? What about at times when they have thoughts of hurting others? What about at times when they are at risk of being hurt by others? Imagine what it might be like for you in such situations. Journal and discuss these thoughts.

Part 3

Having considered what it was or might have been like for you and what it may be like for others in such situations, consider what it may be like for you as a counselor trying to help clients in such situations. What thoughts might run through your mind? What do you think you are likely to feel? How might you be and what might you do? Journal and discuss your thoughts and feelings related to these questions.

INTRODUCTION: THERAPEUTIC RELATIONSHIPS AS A SOURCE OF POWER AND INFLUENCE TO HELP CLIENTS MANAGE CRISES

While we have counseled many clients in situations of imminent danger, we continue to find such situations among the most worrisome. This is partly because imminent danger usually means the possibility of death or other permanent damage. Wrong decisions can be so final. Our worry is also because, while counselors have ethical and legal responsibilities to prevent harm if possible (Corey, Corey, & Callanan, 1998; Wubbolding, 1996), the influence that we have as counselors is or often feels quite limited in such situations.

So, we've worked to figure out what is the most influence that we and other counselors can have in such situations and how to maximize that influence. Perhaps as you have studied *The Heart of Counseling* thus far, you will not be surprised that we find that counselors' primary source of influence to keep clients safe through situations of imminent danger is the therapeutic relationship they form with each client. Even if you wanted to prescribe exact actions for clients in imminent danger to stay safe, you'd probably have no workable way to enforce your directions. In some cases, you might be able to threaten involuntary hospitalization or mental health arrest, but in most situations the case for hospitalization may not be clear and so this would be an empty threat. Besides, your goal should not be just to keep your client safe through an immediately dangerous situation but to keep your client safe so that she can continue to make progress with you in counseling through and beyond the imminent danger. Becoming threatening and controlling around situations of imminent danger would be trading the ongoing healing of your therapeutic relationship for safety through one dangerous situation, while your client may face many more dangerous situations without successful long-term intervention.

We have often made non-self-harm agreements (i.e., an agreement and specific plan for clients to keep their self safe until their next meeting with their counselor) with clients who have strong thoughts of hurting themselves but do not need immediate hospitalization. This is a common practice among counselors. But what would such agreements be based on? Our clients owe us nothing. They do not *have to* do as we ask. We cannot go with them in life to make sure they keep themselves safe. These agreements are based on the therapeutic relationship that each client has with his counselor. The strength of these agreements is the connections made between you and each client.

PRINCIPLES OF MANAGING CLIENT CRISES WITH THERAPEUTIC RELATIONSHIP SKILLS

Self-Responsibility, Dignity, Integrity, and Least-Restrictive Interventions

You will need to balance your responsibility to do all that you reasonably can to keep your client safe with respect for your client's self-responsibility, dignity, and integrity. It may be tempting to just take over and substitute your judgment for your client's. Don't do this unless it is clear that you have no other more reasonable choice. In our experience and that of our friends who are mental health professionals from counseling and related fields, we have yet to come across a situation in which a client could not remain somewhat empowered. So, for example, regarding risk of suicide, there can be various levels of intervention, with the most restrictive being involuntary hospitalization and less restrictive levels including your client agreeing to get rid of the means to hurt self, letting a friend know how bad she is feeling and how strong the thoughts of hurting herself are, and asking that friend to be available to talk to on short notice if the feelings and thoughts escalate.

Acceptance

Remain focused on the person of your client, as well as the crisis of her current situation. It can be tempting to forget the person you are working with and see only her dilemma and your decisions.

Further, if your client has let you know that she is in danger of hurting herself or others—or of being hurt by others—your client may have given you information that was very hard for her to share. It may be information over which she feels shame or embarrassment, or fear that you will then judge her critically or think her stupid, foolish, mean, or manipulative. We have heard mental health professionals say of clients who express strong thoughts of suicide, "Oh, she's only saying that for attention." We see such a nonaccepting counselor attitude as a serious error and probably a symptom of burnout and a sign of the mental health professional also needing care. Such a situation with a client is a highly sensitive moment, a moment in which it is most important that the counselor see and connect with the person who is his client in a dangerous situation. In such situations, you may need to remind yourself to accept the person, regardless of her situation and the decisions she has made that led to her current danger.

Empathy

This is also a moment in which it is crucial to continue the flow of your empathy and redouble your care in expressing it. Your feeling with your client will maintain and deepen your therapeutic relationship with your client in the moment when you need it most. It will also help you understand your client more in that crucial moment and thus make better decisions about how to help your client manage her crisis and enhance your assessment of her level of danger. Your empathy will also help you hear any implied doubts that your client has regarding keeping herself safe or any reluctance to a safety plan that you may be making with this person.

Tell Your Client What's Going On with You

Explain Your Assessment and Decision Process Jeff has realized that he has no choice but to tell clients, whom he thinks may be telling him of imminent danger in their lives, what is going on with him.

> I don't think I have a "poker face." For a counselor, this is not a problem and may be a great asset. So, I think that if I am distracted in listening because I am also beginning to assess for potential danger, this distraction would be apparent. I'd much rather tell my client what my distraction is

than have my client sense that it's there but not know what it is. If my client tried to guess why I seemed distracted, he might very well assume critical thoughts, boredom, and so on. So, once I realize that I am distracted, I find a moment to say so. I usually say something like, "I realize as I'm listening that I'm also distracted. When you told me how tired of it all you were and that you thought you just couldn't go on, I took this to mean you are thinking of suicide. Now I find myself trying to figure out if you are safe, while I also try to listen to the rest of your experience."

Following a statement like that to an adult or adolescent client (the need for genuineness with children is also quite true, but we would not confuse a child client with our adult decision process), we find that clients usually help us or other counselors with our struggle and begin to tell how they assess their safety. If not, we may ask for what we want or need to know. Additionally, we might further explain, saying something like, "Right now, I'm worried for you. I very much want you to be safe and I have a responsibility for your safety." Look for examples of such interactions in the case examples we provide later in this chapter.

Carefully State Your Feelings Because it is always true, it is easy for us to tell clients in dangerous situations that we care for them, that we want them to live and be safe and well. We may also tell them what we feel in response to their situation, such as worry, concern, or fear (Note: This is *our* feeling, not empathy). While we do sometimes use this statement of our feelings to partially justify our request for them to help us with our assessment or to plan with us for their safety, we try to make this statement of what we feel in response to their situation flatly, meaning that we are not implying that they caused our feelings or are obligated to do anything about it but that we do want them to know how we are affected by their situation.

It is also often easily true for us, so when it is, we are able to tell clients our hopes for them. This usually means we want them to live and be safe because we want them to and believe that they can make progress in counseling. We might make a statement like the following: "While everyone's situation is unique, I know that persons with situations that seem similar to yours have come to improve through counseling. So, I am hopeful for you. I believe you can work through this, especially with your hard work in counseling."

Remember to Reflect

Throughout the process of your assessing and helping plan for safety, you should not stop therapeutic listening and frequent reflection. This remains the core method for you to understand and connect with your client and for your client to know that you are understanding and connecting with her. You are layering your skills in assessing and planning for safety over your skills in therapeutic listening, genuine empathy, and UPR, rather than substituting assessment and planning for these skills.

So for example, if one of us has told a client that we have begun listening *and* assessing, that client might respond, "Oh, I'm fine. Don't worry." We would reflect the content or emotion that we heard in this statement, rather than switching immediately to assessment. If we felt some irritation communicated with our client's statement, we might respond, "You want me to know you're fine and you feel a little irritated that I asked." Or, if we felt surprise with our client's statement, we might respond, "You're surprised that that was what I thought. Seems like it must not have been what you meant."

Or if, for another example, after expressing a feeling of worry or concern in response to our client's situation, our client responds, "Oh, I didn't want you to worry," and we hear they're upset over this, we might respond, "I see you hadn't wanted to trouble me." We would make these reflections before or instead of continuing assessment or planning. However, if we still see a need for assessment or planning for safety, we would also continue that process.

Help Your Clients Make *Their* Plan

When you do help your clients plan for safety, a key to remember is that you are helping your clients plan for *their* safety, meaning you are helping your clients make plans that they will own and implement—the plans will not be yours. It may be tempting, especially if your time is limited—which it probably always will be—to take over and make *your* plan for your client's safety. However, this may produce a plan that your client does not have any personal ownership or investment in, one that your client has not thought through in terms of how implementing this plan may work and feel in her life, and thus a plan that your client is unlikely to actually implement.

Plan Specifically

Help your client anticipate any reasonably likely pitfalls. Make the plan as specific as possible and suggest any reasonable difficulties that occur to you around each action within the plan. For example, if your client is planning to call and let someone know when he has strong thoughts of hurting himself, he will likely need to decide specifically who to contact and may need a backup resource or two in case the first contact person is not there. Then, if it occurs to you that he has had great difficulty paying his bills and that his utilities have been shut off at times, you may need to ask if he is sure to have access to a phone at times that the two of you know are high risk.

Err on the Side of Caution

Always err on the side of caution in such situations. For example, in our work we have sometimes felt concern that our intent on specificity and anticipating possible pitfalls may seem condescending. Actually it seems to have been perceived as caring, but even if it were perceived as condescending, we'd rather take that risk and be more sure of our client's safety. We can address any apparent perception of condescension within our therapeutic relationship, through therapeutic listening, and careful, deep empathy and UPR.

Additionally, if you think your client is implying a dangerous situation to you but are not sure, tentatively reflect or ask this. You won't plant the idea in your clients' mind. More likely you'll deepen your therapeutic relationship through your expression of your honest concern. If you misunderstood and had it really wrong, accept and acknowledge that correction. However, if your nagging doubt seems really to be what your empathy is communicating to you, then you may need to reassert your concern. It may be that your empathy is picking up an attitude that your client is as yet denying (perhaps in awareness or not). Be as patient as possible in allowing your clients to come to their realizations of concern. More experienced counselors have had practice with clients revealing themselves over time and have discovered that rushing to an insight is not usually productive. Yet, if you continue to see serious reasons for concern and your client does not, respectfully assert your concerns.

Say the Words

If you think that you and your client are talking about suicide, go ahead and use the word *suicide,* or phrases like "killing yourself," or, at least, "seriously hurting yourself," whichever is most true of what you are actually talking about. This is not meant to shock but to be specific and clear and truthful about what you are talking about, without undue hesitation. You can, of course, say the words with sensitivity and caring.

Respond to the Possible Communication
of a Dangerous Situation As Soon As Possible

If you think your client has stated or implied a potentially dangerous situation in her communication to you, don't wait until late in your time together to respond to it. For example, if your client has begun to cry and have great difficulty explaining a difficult situation to you and you've heard her say, at least once, maybe twice, something that sounded to you like giving up on life, you may respond to the strong emotions first but get back to what sounded to you like giving up on life as soon as you can while also attending to her emotions in that moment. While beginning to cry, a client with a serious illness might say, "I'm just so tired of being sick. I just can't stand it." Then a little later, she might say, "I just don't know how much more of this I can take." So you might tenderly, empathically respond, "Wow. I'm beginning to get at least an idea of how hard this is for you. I also want you to know that as you spoke, I began to get the idea that you have thoughts of seriously hurting yourself, of ending this suffering that way." In such a situation, if your client does not then elaborate on thoughts of hurting herself, ask about this directly. In such a situation, your question could be phrased, "Shondra, in all this pain, are you thinking of hurting yourself?" (We inserted a fake name in this example because we couldn't imagine not using our client's name in such a personal moment.)

A point that we want to be sure is clear to you from this subsection is that you should not wait until you are sure from your clients' statements that they are in imminent danger. Instead, it is better to respond when you only have a suspicion. In that way, you may open that part of your counseling session while you still have time to listen, maintain your empathic connection, *and* assess for danger and plan for safety.

However, it may well happen that your client first gives you reason to suspect imminent danger near the end of a meeting. If that happens, we see this as one of the few legitimate reasons for extending the time of a session and even being late for another.

CONSIDERATION OF
ASSESSMENT FACTORS

We offer a case illustration of our principles of managing client crises with therapeutic relationships skills. But before reading the case illustration, it will help you to review the factors we normally consider in assessing such risks and our thoughts on non-self-harm agreements.

The following are the factors we would normally consider in assessing or helping clients determine their risk of suicide or seriously hurting themselves. You should of course use your own sound judgment when making such assessments. You may not need to consider all these factors with each client, and others may occur to you as important in some clients' situations.

A Plan

The factor we often consider first is whether or not a client who seems to have thoughts of suicide has a plan to hurt himself. A client may have only vague thoughts of killing himself, like wishing for the pain to be over, wondering what it's like to be dead, or imagining that others in his life will miss him or feel guilty, but has never thought of *how* he might suicide. In that case, depending on other factors, we may decide that this person is only a minimal risk of hurting himself and mostly needs his counselor to listen carefully to his thoughts and feelings; to ask about these thoughts, any plan, and his determination of risk; and then watch for any change or worsening. On the other hand, if a client has a specific plan of how he would kill himself, this would concern us much more. It may indicate both how much he has thought about it and how close he is to suicide.

When we are following up on what sounded to us like thoughts of suicide, after reflecting that we thought our client may have been talking about suicide and listening to her response to this reflection, we might phrase a question something like, "So, when you think about death and the release that it might bring, I wonder, do you ever have specific thoughts of hurting yourself?" Or, if a client already tried to assure us that while she does think about it, she would never actually attempt suicide, but we're still worried, we might reflect and ask something like, "I hear you assuring me that you'd never actually hurt yourself, but I still worry. So, would you tell me, when the thoughts cross your mind, do you ever think of actual ways that you might hurt yourself?" It is important to ask such questions respectfully and to seek specific, definite answers. It is also important to ask with an understandable and true level of concern but also with acceptance and without panic. In such situations, it makes perfect sense for you to be concerned, but panic can leave a client thinking you don't trust or respect her, and that can shut down her communication to you.

Lethality of the Plan

Secondary to having a plan, we would want to know if the plan seems lethal or likely to do serious damage. However, most plans seem serious or at least potentially lethal to us. For example, if a person has a plan in mind to crash her car into a tree, it might not be lethal, but there is a chance that it would do serious harm, so we would still take further action to become more sure of her safety.

A common plan for suicide would be overdosing on medication. Because we would usually not know what danger an overdose with a specific medication may pose, we just assume it is serious or lethal.

A Means

If a client has a plan that might be serious or lethal, then we'd want to know if the client has the means to implement the plan. If a client says, "Yeah, that's how bad it is. I've thought a lot about it. I've even thought that if I ever did it, I'd use a gun. I'd want no chance that I was injured but not dead." To which we might respond, "You're hurting badly enough to think of using a gun. You'd want to be sure to actually die." Then, assessing the availability of the means, "Look, I see your point about the certainty of using a gun, and I'm trying to figure out if you're safe right now. Do you have a gun or access to one?" Then, if a client has access to a gun or other lethal means, we'd see him as being in a high level-of-risk environment and take every reasonable action to provide for his safety. If the client instead tells us that he does not have a gun, know anybody who does, or indicate any plans to get one, we'd still take his situation seriously but would have a possible indication that he does not yet intend suicide.

Preventive Factors

Whether or not a client has a plan and a means, we may also want to know what has kept that person from suicide before, if she'd let us know it has been on her mind. For example, if a client tells us that she wouldn't kill herself because she knows it is a sin, and we know too that she has been troubled by her church and has been greatly questioning her spiritual beliefs, we may see the precarious nature of this preventive factor. If, on the other hand, a client tells us that he wouldn't do it because he just couldn't hurt his mother that way, and we know how much he cares for and values the strength of his relationship with his mother, we may have great confidence in this preventive factor, even while we still see the seriousness of his situation.

Future Orientation

Whether or not a client seems oriented toward her future may help us to be more or less certain of an assessment of her safety and of potential plans to keep her safe. For example, if a client of whom you are concerned for her safety seems to genuinely want to leave her session on time in order to not be late for class, which would affect her grade, then that might be an understanding of her and her situation that could help you see that she is future oriented and perhaps not in immediate risk. If, on the other hand, that same client did not seem concerned with making it to work or class, that might be more reason to worry and to even more carefully assess and help plan for her safety. Our general thinking here is that if a person seems to have stopped planning to live, then we might assume that that person has knowingly or unknowingly already decided on suicide.

A Sudden Change or Switch

If a client has had a sudden change in his way of being, this may cause us greater concern. For example, if a client, who did care very much about his work, school, or children, suddenly is much less concerned, we might worry that this means he has given up on life. If a client has always been very driven and inspired in his actions but this changes to a suddenly easygoing demeanor, if we were already concerned with his thoughts of suicide or feelings of depression, we may become more concerned. Even if a client has usually been easily irritated and suddenly no longer seems to care about the same irritations and we were already concerned with his thoughts of suicide, we may become more concerned. You might reflect that possible change, withsomething like, "Look, John, I know how down you've felt and that you've even had thoughts of suicide. So this sudden change has me thinking, worrying that maybe you've given up on life."

Previous Attempts

We would often want to know if a client whom we are concerned about committing suicide has ever attempted suicide before. If we know that a client has attempted suicide before, she may help us and help herself assess her current risk by considering how her current situation is different from the previous time. For example, does she feel as bad as before, does she have a better support system now, does she have different things in her life that may keep her from attempting it?

Lowered Inhibitions and Impulse Control

If we do not already know, we would also want to know if a client whom we are concerned about committing suicide is engaging in behaviors that seem to lower inhibitions or impulse control. Unless that client is a very new client, chances are we already know about any significant alcohol or drug use, as this is information that our clients have commonly shared with us. If we don't know but we are concerned, we would ask. For example, we might say, "I don't remember your ever having mentioned your using drugs or alcohol, but I want to ask you about this because I know that alcohol and most drugs tend to lower people's inhibitions and impulse control. If this client hasn't already guessed where this was going and interrupted with the information, we might add, "Is alcohol or drug use a part of your life, something you ever do?" Then, if this client's answer is not already specific, we would ask him to help us specifically understand his alcohol or drug use.

Ability to Guarantee Safety

Especially when planning for safety, if a client can convincingly guarantee her safety, we would worry less about her safety but continue to take it seriously, of course. By guaranteeing safety, we mean things like agreeing to give up any means to complete a suicide thought or plan, or agreeing to call a specific person in case of strong thoughts of suicide. Especially in determining this factor, as well as in others such as future orientation, thoughts of a plan, and preventive factors, your therapeutic relationship and especially your empathy are key. If a client is just telling you she agrees to a plan for her safety to satisfy your worries and you are attending to your client with careful, strong empathy, you will notice her expression of reluctance or reticence. If you think you hear something like this, you should of course reflect that. In such a situation, you might reflect something like, "So you're telling me that you will call the crisis line if you have strong thoughts of hurting yourself, but you seem to be only reluctantly agreeing to do this."

And Finally . . .

Remember, while we've listed factors that we would likely want to know about in assessing for safety, we advise you not to switch into assessment, question, and answer mode. When you are assisting a client who may be in imminent danger, it is no time to abandon your therapeutic relationship skills. Don't establish a pattern of your repeating questions. Remember to reflect.

NON-SELF-HARM AGREEMENT/SAFETY PLAN

For a client about whom you are concerned over his thoughts of suicide, consider the previous assessment factors in making a safety plan. Then, make the plan as specific and concrete as possible while remaining focused on the person of your client, his feelings, thoughts, actions, and way of being, and your therapeutic relationship with him.

A Good Safety Plan Is Time Specific

Such plans are usually time specific. To ask a client who is really hurting to promise to take actions to keep herself safe forever, or for a time that seems like forever to her in light of her pain, is too much to ask. So, we usually just plan with such clients for between the time we are making the plan and the next time we meet. If waiting for a traditional weekly appointment seems too long, we would plan, if able, to meet with her again in just a few days.

Relate the Safety Plan to Avoiding Elements within Your Client's Suicide Plan or Thoughts

If your client has a plan or specific thoughts of how he would suicide, it is important for him to avoid the elements of that plan or those thoughts, if possible. So, for example, if your client has thought of getting a gun and shooting himself but does not have ready access to a gun, then a part of your plan with your client may be for him to take no action toward getting a gun between that time and the next time that you meet. Or, if your client had always thought that he would go to his family's cabin alone and take an overdose but has not acquired any medications for overdose, he should plan to avoid having time alone at his family's cabin in order to not be so tempted and agree not to acquire any medications for potential overdose.

Get Rid of the Means

If your client has a means to implement a suicide plan, see if she is willing to get rid of that means. (If your client is unwilling to get rid of that means, it may mean she is not really committed to making and implementing a safety plan.) Examples of getting rid of the means might be dumping out unneeded and potentially dangerous medications, or disabling a gun. If the client needs the medication or is unwilling or unable to disable the gun, an alternative might be to have a trusted love one take possession of the medication (to give back in small amounts at a time, as needed) or take possession of the gun.

Avoid Lowered Inhibitions and Impulse Control

If you think of anything your client might do that would lower her inhibitions from suicide or lower her impulse control of taking impulsive suicidal action, then plan with her to avoid these actions. Often in such situations, we would be concerned with a client's alcohol or drug use because alcohol and drug use are commonly known to lower inhibitions and impulse control. This can be difficult to plan for, because if your client is seriously addicted, she may not be able to immediately give up drug use. If it seems impossible for your client to give up drug use, a more restrictive plan may be required. For example, your client may need to be hospitalized for safety or to contact a loved one to watch over her during the crisis time.

There are a few ways we have found to make safety plans in which our clients remained more fully self-responsible (i.e., not hospitalized or given over to the care of a loved one), even when possibly addicted to alcohol or drug use. For example, with some clients we have planned for only half a week at a time. To some clients who drank or used drugs frequently and found it difficult to stop, it became much more reasonable to plan for staying away from alcohol and drugs for just a few days. Other clients who knew they would have great social pressure to go out and drink or use drugs have decided to tell some of their friends whom they thought might pressure them either that they have been so down that they've actually considered suicide, so they are going to stay away from alcohol and drugs for at least a few days, or short of mentioning suicide, have told them that they just felt so down lately that they were going to stay away from alcohol or drugs for a few days. Other clients have decided they would be all right if they just drank in moderation and so have asked a friend whom they knew was able to drink in moderation or not at all to watch over them and help them keep to moderation. As these requests require telling friends of their thoughts of suicide or at least of how down they have felt, they seem to speak to the clients' honesty in assuring their sincerity.

Prevent Harm by Contacting Someone Immediately

Almost every safety plan we have made that did not require hospitalization or giving one's self over to the care of a loved one included a specific plan of whom the client would contact when he had strong thoughts of hurting himself. This can include a friend or loved one, whomever you and your client decide will be available and respond reasonably. This often includes a suicide hotline or a night emergency number of your setting, or another agency that is able to respond around the clock. This usually includes more than one option (e.g., first-choice friend to contact, second-choice friend or family member to contact, then, if neither of these are available, the community's crisis hotline). The key here is to get in contact with someone who can help before the thoughts get so strong that they are hard to stop and your client acts to hurt himself on impulse. So, you may help your client understand that this is not for just a passing thought but does mean that he will call for help just as soon as he notices that his thoughts of self-harm seem to be persisting.

Don't Complete Assessment and Plan Alone   Serious thoughts of suicide are a reason to break confidentiality. If you work in a setting with other counselors, it seems reasonable for you to discuss it with your client first, then invite in a coworker to review your and your client's decisions and plan before your client leaves. If you work at a setting where you are the only counselor, you may want to bring in another professional, such as a principal or teacher.

Work Slowly and Carefully

Especially if you cannot find another professional to consult with before your client leaves, work slowly and carefully. Even when you are well practiced at these skills, it will be difficult to remember everything at once. So, pause to think when you need to, refer to notes if it is helpful, and remember that you are planning *with* your client, not *for* your client. Remember to have your client share in the process and to consider her self-assessment and her thoughts and feelings regarding the quality of the safety plan and all its pieces.

And Finally . . .

Remember, the strength of your client's safety plan is the strength of the therapeutic relationship that you establish and maintain. Don't get caught up in planning and forget to attend to your client as a person, to listen therapeutically and respond with genuine, heartfelt empathy and UPR.

A CASE EXAMPLE OF A CLIENT WITH MILD SUICIDAL IDEATION

It will likely be helpful for you to read an example of a counselor using, developing, and maintaining a therapeutic relationship while also helping to assess and plan for the safety of a client who may be in imminent danger of suicide. While no two situations are alike, the following example is a realistic, reasonably typical example of such a situation in which the client can leave the session most fully self-responsible (i.e., not hospitalized or turned over to the care of a loved one). This example also employs most of the assessment factors and elements of a safety plan and, most importantly, our principles of managing client crises with therapeutic relationship skills as they apply to this client's situation. While reading, look for evidence of the counselor using, building, and maintaining a therapeutic relationship while helping her client manage this situation of imminent danger.

[Picking up midway through the third session.]

Client: [In exasperated tone.] I'm just so tired of it. I wish I could sleep forever. I've got so much on my mind, so much hanging over me.

Counselor: [Matching tone.] You just feel overwhelmed.

Client: Yeah, overwhelmed and tired of it.

Counselor: [Reflecting her immediate statement and going back to one that she saw as important to her client earlier.] Tired of it, too. A minute ago, I noticed you said you wished to just sleep forever. I took that as a big statement of just how hard this is for you. I took it to mean you wish to die.

Client: [A little shocked to hear this said outright.] Well . . . [stretches this word out, then pauses and looks down to think, then responds more definitely] no, I wouldn't do that [pauses and looks down again, briefly]. No, I wouldn't do that.

Counselor: [Accepting her definitive statements *and* the uncertainty she expressed in making them. Also having some difficulty getting the harsh words out.] So, you know that you wouldn't hurt yourself, well, kill yourself. You know that, but it seems you have thought about it and feel some doubt.

Client: [Pauses before speaking. Her way of speaking changes from sounding definite at first, to halting, to sounding overwhelmed and afraid again, and worried by the end of her statement.] Yeah, I just couldn't do it [pause], but I do think about it [pause], well really I think a lot about it. The worse this gets, the more it's on my mind.

Counselor: [Continuing in her client's tone and beginning to realize that her client is concerned for and assessing her own safety.] Oh, and it's on your mind enough that this is troubling you; these thoughts of suicide are worrying you.

Client: Oh, it's silly. I'd never have the nerve to do it. I hate pain.

Counselor: [Starting too quickly, beginning to think of assessing.] So, that's one of the things that—[stops somewhat abruptly, realizing that she was about to be only going through the motions of therapeutic listening]. Gina, I realize that I'm concerned for your safety. While I'm still wanting to understand your experience, I also find my mind trying to think through just how safe you are in this very difficult time. I know you said that one thing that keeps you from it is the pain. I also heard you dismiss it, saying it was silly for you to even think about it.

Client: Yeah, I'm just a big baby. I don't even deserve this. I'm just feeling sorry for myself.

Counselor: [Reflecting what she sees, with empathy in her tone.] You have no patience for yourself with these thoughts.

Client: [Rising frustration in tone.] Yes, it's wrong. It's a sin. God, I'm so weak!

Counselor: You see it as an absolute wrong and yourself as weak for even thinking of suicide.

Client: [Pauses for a long-seeming half minute, then shrugs at counselor, as if to say, "I give up, I have nothing more to say, what now?"]

Counselor: [Reflecting with tender empathy, pausing, then questioning.] You're so tired of this [pauses as client looks down]. Gina, as I think about your safety from suicide, I notice that one thing that keeps you from it is the pain; another might be that you know that it's a sin [client nods at both, tearing a little]. If you will allow me to ask [she looks up, indicating consent], I wonder if, even while you'd not likely ever even to attempt suicide, when it's been on your mind, you'd thought of how?

Client: [Disliking the situation but not the question.] Oh god, how did I get to this place? [short pause] Well, you know I have a lot of pain pills around. Last weekend I lay there one night alone, lonely, feeling sorry for myself, of course [this last part said with at least mild self-disgust], and I thought, "You know, I could use a little wine anyway to help me sleep." You know I almost never drink anymore. It just puts me to sleep. I've had the same bottle in the fridge for weeks. Anyway, I thought of it then. And I thought, "I've got half-bottles left of a couple of different kinds of pain pills. If I took them all with the wine, it just might kill me. Then this would all be over and my sorry excuse for a family would really be sorry then." But you know, I chickened out. God, I can't even do that!

Counselor: [Genuinely moved.] Ungh. You are really hurting with this. You're down from your situation, down on yourself for thinking of suicide, and down on yourself for not going through with it.

Client: [Crying, trying to stop crying. Whispers.] I'm so sorry.

Counselor: [Whispering a reflection in matching tone, then shifting to assessment for another moment.] So sorry [pause]. [Speaks tenderly, clearly and carefully] Gina, I want us to plan for your safety. Will you work with me on that? [Gina nods her assent.] I want to ask some questions and suggest ways to keep you safe. [Gina looks up and straightens, indicating her readiness.] [Counselor realizes the time.] Oh, I realize we would normally have only five minutes left. I think this is a very important and unusual situation for you. It's important to me to be sure you will be safe. I'm going to call our secretary to ask her to ask my next appointment to wait. Can you stay another half-hour? [Gina agrees and the counselor calls to arrange the additional time.]

Client: [Having waited through this short phone call, Gina smiles when her counselor's attention returns to her.] Thank you. I'm sorry to be so much trouble.

Counselor: You see yourself as trouble. I want to do this [Counselor states this tenderly but without undue emotion, as it is the truth of her reaction to her client]. [Counselor takes a short pause to think] Besides the pills, have you ever thought of other ways of hurting yourself, of suicide?

Client: No, really I won't do it. Gee, what if it didn't work? How embarrassing!

Counselor: Oh, so you know you won't and there's another strong reason not to attempt suicide—you'd feel terribly embarrassed for others to know. [Gina shrugs.] I'm wondering if you need all the pain pills?

Client: Well, maybe not. The one bottle is nearly empty. I take stronger ones now. And the other I don't take anymore, 'cause when I started them, they made me sicker. I could just get rid of those bottles.

Counselor: So, it sounds like you'd be willing to get rid of them completely.

Client: Then, with the ones I use, I can just have the prescription filled in bottles with fewer of them, 10 at a time. Smaller bottles are usual, anyway. They just made up larger bottles for me.

Counselor: You seem okay with this idea. [Gina nods her agreement.] I'm relieved to hear this is possible.

Client: Yeah, I'll be okay then.

Counselor: You say that with much less hesitation. You're becoming more sure now.

Client: Yes, I do appreciate your concern.

Counselor: So that has been helpful to you. [Gina nods and smiles.] If I may, I want to ask a couple more things. First, I know that alcohol can lower inhibitions and impulse control, but I also remember that you never really drank to excess, right? [Gina clearly affirms this in her gestures.] Even so, I worry that you might get even a little tipsy and increase your risk.

Client: No, half a glass just helps me sleep some nights. When I have that half a glass, then I just get to sleep right away. [Gina says this with clear confidence.]

Counselor: Okay, my other thought is that normally with a person suffering so much as you seem to be, I might ask that you call someone, perhaps the suicide/crisis hotline when you have strong thoughts of hurting yourself.

Client: I think I can do that. I've thought of telling my minister, 'cause I wondered if, you know, for spiritual reasons, what it would mean, but I just didn't want her to know. But this crisis line, it's private right? No one has to know it's me, 'cause I feel silly enough already?

Counselor: Yes, it can be completely private. You seem quite sure of these steps we are planning.

Client: Yes, I'm glad to have a plan.

Counselor: Okay. Well, I'm afraid I'll seem a little condescending, but if you'll let me, I'd like to review to make sure we have a plan for your safety until the next time we meet. [Gina says okay and looks up, paying attention.] So, you'll get rid of the pain pills that you don't need and keep only small amounts of the ones you do use. [Gina nods to say, "Okay, got it."] And you will get rid of the useless ones as soon as you get home. [Gina smiles a little at the quizzing and says yes.] Then if you have strong thoughts of hurting yourself, by strong I mean that they continue for more than a passing thought and you begin to get that anxious feeling [Gina nods, meaning both that she agrees and that she remembers that anxious feeling that her counselor refers to. She has described it before], you will call the crisis hotline and talk to the person there about what you are feeling until the thoughts have passed and you can go on. [She nods her agreement again.] And you can keep this plan until we meet next week.

Client: Yes, really I'm fine, well, better now. I'm glad for your help and for having helped me think this through.

Counselor: [Smiling.] Okay, thanks for indulging my need to be sure. You are important to me. I'll get you a pamphlet about the crisis hotline on the way out and we'll schedule for next week.

THE ISSUE OF HOSPITALIZATION

Situations where clients have had to be hospitalized, as in the case example we give following this section, have been vastly more rare than situations in which we have been able to successfully plan for clients' safety in our work. This has also been true of our students' and friends' works. But even while such situations are rare, it remains critical for you to think through and develop your readiness to handle such situations, as they are life-or-death situations.

Know Your Local Laws, Guidelines, and Procedures

Every community has laws and procedures that guide the behavior of mental health professionals regarding hospitalizing persons for their own safety. Each counselor should make herself familiar with the laws and procedures of her community and the rules or guidelines and procedures of her work setting.

When to Seek Hospitalization

Usually, a client would be helped into the hospital when he is unable to guarantee his safety in a way that satisfies his counselor. Generally this means that a client will not agree to a non-self-harm plan with confidence. For example, if a client has a plan for how he would suicide and has a means but will not agree to give up the means and/or avoid other risky behaviors, then his counselor would have strong reason to doubt any guarantee of safety and would likely urge or insist on hospitalization in order to insure his immediate safety.

Maintaining Maximal Client Self-Responsibility

We used the words "urge or insist on hospitalization" because, even in cases requiring hospitalization, we believe in keeping clients maximally empowered and in charge of their self and decisions. While with help a counselor probably could force a client into the hospital for her own safety, so far in our works this has not been necessary, and in the works of friends who are mental health professionals, it has almost never been necessary. However, we and our friends have helped many persons decide on hospitalization for their own safety.

Responding with Empathy to Clarify Intent

We also used the phrase "unable to guarantee their safety in a way that *satisfies* his counselor" intentionally. An example of what we mean by "satisfies" could be a situation in which a client is stating his or her agreement with an aspect of a non-self-harm plan but seems to be communicating something more at the same time. If one of us had a client who agreed to a non-self-harm plan while sounding annoyed and saying, "Okay, okay, I'll do it." The one of us who is counselor to that person would likely respond with something like, "So you're telling me that you will, but you sound annoyed, like I've just nagged you into this, but you don't really agree." To which a client might respond, "[calming and becoming less argumentative in response to our UPR inherent in the tone of our response] Oh, it's not that, really. I just hate going to all that trouble. But I guess that's where I'm at right now." Then, we might proceed with greater confidence to conclude a non–self-harm agreement.

If, on the other hand, a different client started from that same annoyed statement, "Okay, okay, I'll do it," and the same accepting response, to then say, "Well hey, I know the game. I know what you gotta hear so I can go home," then that response would have confirmed our doubts, and we would seek to have this person choose hospitalization in order to stay safe in the immediate future.

Counselor Responsibility

However, we wish to carefully reiterate that while we think in terms of "urging and insisting" and clients "choosing hospitalization," we know that counselors and other mental health professionals do bear an ethical and legal responsibility for their clients' safety (Corey, Corey, & Callanan, 1998; Wubbolding, 1996). So, we see care for clients in such situations as a balance between maintaining maximal client empowerment and self-responsibility with the responsibility that mental health professionals have to protect clients' immediate safety. We illustrate this balance in the case example that follows shortly.

The Issue of Paying for Hospitalization

Before presenting the following the case example, we wish to add one more thought. In helping clients choose and plan for hospitalization to maintain their immediate safety, we usually also discuss with them how they will pay for it. We think of this as only being considerate during a difficult time; we don't want them to be surprised with the issue at the hospital. So, as much as possible, we help them think through whether or not their insurance will pay. Often times we've been able to call the hospital we were considering and have their receptionist direct us to help with this. Through similar actions, we've also been able to help the client start planning for payment arrangements when she does not have insurance. And finally, for young adults who will need their parents or parents' insurance to pay, we've been able to make this initial parent contact for or with them.

AN EXAMPLE WITH A CLIENT EXPERIENCING STRONG SUICIDAL IDEATION

Please note: We have changed this case example so that it is similar to the situations of numerous clients we have served but unlike any particular person. However, we tell it in the first person, as if straight from memory, in order to make telling it less cumbersome. We've also created a fictional name for this same reason.

A teacher brought Wanda to counseling. Wanda had completed a short story assignment in which she depicted herself sitting in her garage with a concoction of dangerous chemicals mixed in a jug and intending to gulp them down. The teacher was obviously and naturally worried that the story was true. She explained this to me, in my office with Wanda, and then left us to talk privately.

I had worked with Wanda the year before. I knew her to be a bright and driven student, hoping to specialize in chemistry and maybe one day study medicine. While she'd not yet articulated it clearly to me, I knew that she felt a great pressure to succeed academically for her family and doubted her abilities. When we worked together before, Wanda had been referred by a different teacher for apparent test anxiety. Wanda had not wanted to engage in personal counseling but only wanted to learn skills that she could use to manage her test anxiety. So, I had helped her learn cognitive-behavioral emotional management skills (Beck & Weishaar, 1989; Ellis, 1989) and encouraged her to continue in counseling to explore herself through her feelings and in that way make changes that would complement, augment, and enhance the skills she had requested. I respected her decision, and while I had known she felt strong, worrisome pressure, I had not been concerned for her safety. Now that she was brought back to counseling, I was glad that she had apparently used her short story to let her teacher know of her danger.

The following dialog picks up early in our session, after the teacher had left.

Counselor: So the scene in your short story is real. [I reflected this calmly, as a fact, even though I was also immediately concerned for her safety.]

Client: [Looking me straight in the eye, with a thoroughly serious, determined look.] Yes, some night soon, I will do it.

Counselor: And you are absolute about that. No "might," only "will."

Client: [States no response but looks sometimes into my eyes and sometimes down, seeming determined to hold her ground on her intent to kill herself.]

Counselor: Wanda, I'm deeply concerned for you. I want you to live. I want you to give us a chance to work together to discover changes that will help you want to live again.

Client: [Continues to stare or look down with a determined look on her face.]

Counselor: Wanda, you seem so determined to kill yourself. I imagine you must be deeply hurting over something or some things in your life.

Client: [Speaking in measured tones but with apparent irritation.] I just want to die. It's my decision and I want to do it.

Counselor: Ughmph! [A genuine sound that bubbled out of me, seeming to indicate that I'd been hit hard by what she'd just said.] [Continuing after a pause of a second or two] While I want to respect your wishes, I also want you to live, and I have a responsibility for your safety. While I'd like to hear any concerns you might wish to express, I hear clearly that you are not safe, that you wish to die, and that you have a ready means to kill yourself.

Client: [Simply nods, then waits for me to continue after a short pause.]

Counselor: Still, I'd like to ask a couple more questions to see if there is a way that we can keep you safe.

Client: [Shrugs, as if to say, "Go ahead."]

Counselor: May I ask that you get rid of the chemicals that you would use to kill yourself?

Client: Yes, but I can always get more. And besides, it's all stuff we need around the house, antifreeze, bleach, gas, you know.

Counselor: Okay, so you don't see that idea as useful. [After a short pause] I notice that something has kept you from killing yourself so far, that at least you've paused to think about it.

Client: Oh, I only need time to get up my nerve. I'm afraid, but I'll build up my nerve soon.

Counselor: Yes, you have me convinced. [After a long pause to collect my thoughts] Wanda, you seem absolutely unsafe on your own right now. [The look on her face continues to confirm this.] So, I want to find a way that others can be involved in keeping you safe in the immediate future. Two ideas occur to me. We might be able to contact your parents and see how they can help, or we may have you admitted to the hospital in order for the professionals there to keep you safe and evaluate your ongoing safety.

Client: [Starts to get up, seemingly shocked, outraged, raising her voice.] There ain't no way you're putting me in the hospital!

Counselor: [Interrupting quietly but firmly, remaining seated.] Wanda, please stay. [Wanda sits back down.] I want to find a way to keep you safe.

Client: Call my parents then, but no hospital.

Counselor: So that's the idea that seems better to you. You'd be okay with that. You seem to think that might work.

Client: [Shrugs, apparently still angry.]

Counselor: You seem so angry. [Wanda gives no further response.] I assume then that you trust them to help you stay safe?

Client: I guess.

Counselor: Wanda, I'll call them, but you're not exactly inspiring my confidence.

At this point, I asked for and Wanda gave me her stepfather's phone number and her consent to call her parents. Before calling, I checked if there were things that she specifically did and did not want me to say. She didn't stipulate anything. Fortunately, I was able to get him on the phone right away. I spoke to him briefly from my office phone. He called Wanda's mother/his wife, then called us back at a speaker phone in a conference room that Wanda and I moved to so that all three could take part in the conversation.

He concurred that he wanted Wanda to come home instead of going to the hospital. He said that he and her mother had decided that she gets too stressed over school and so should stay home a few days. Picking up in the middle, my conversation with them went something like the following:

Counselor: [regarding Wanda staying at home for a few days] So, I guess you or your wife would be there with her to watch to see that she is safe?

Client's father: Well, no. We'll have to go back to work [he begins to tell of their jobs and the demands on their time].

Counselor: I'm not sure you understand my concern. Wanda means to kill herself. She has a means to do this and she means to do it.

Client's father: Is this true, Wanda?

Client: [after a pause] Yes. I don't know. [another pause] Yes, it is true.

Client's father: [to Wanda] Why would you do this? You know we'd do anything for you.

Client: [interrupting] Do anything? You're never there!

[They argue for a short time and then I interrupt.]

Counselor: Mr. Jones, I want to admit Wanda to _____ hospital, where the professional staff can keep her safe while they evaluate and help plan for her ongoing safety and wellness.

Client's father: Yes, but that's not necessary. Please, we'll just come get her. I'll be there in an hour.

Counselor: I see that that's what you want, how you would help, but right now, I am also responsible for her safety. I worry that it may take time for you or your wife to clear your time, and I worry that there is some discord between you that may make it hard for you to help right away and there may not be time.

Client's father: Wanda? [she doesn't answer]. Are you there?

Client: Yes, I'm here.

Client's father: Is it true, baby, 'cause you just can't do this. You know we got a lot going on right now [Wanda sighs, slumps, and covers her face].

Counselor: [speaking calmly, carefully, and respectfully] Mr. Jones, there are a couple of choices to make. If you want her admitted to _____, rather than _____ hospital, you will need insurance or a way to guarantee payment. Can you do that?

Client's father: Well, yes, if you can't just let us handle this at home.

Counselor: So, that's still what you want, but I hear that you're willing to go along with my concerns and request that she be admitted to the hospital. [after a pause, in case he wanted to say more] If the police drive her to the hospital, they will have to take her in handcuffs. Would you rather drive her?

Client's father: Well, yes, if you really think all this is necessary.

Client: Okay, look, I won't do it. [rolling her eyes a little as she speaks] I'll get rid of the chemicals. I won't do anything to hurt myself.

Counselor: You're telling me that, but I gather it's just to ease the difficulty for your family and yourself right now.

Client: No, I mean it.

Counselor: So, you really mean it. I know that you want to get us out of this dilemma, but my worry is that as sure as you were of your decision to kill yourself just a little while ago, you might change your mind again [she says nothing to this but looks at me in anger]. You seem furious with me now.

Client's father: [after a pause] Mr. Cochran, I agree with you. I'm leaving work in just a few minutes and I'll be there to drive her to the hospital.

He did come. While we waited, Wanda and I called the hospital and spoke with an intake counselor, who concurred with my rationale for the hospitalization and was able to prepare for her arrival. Later, as he had agreed to do, her father called and left me a message that she had arrived at the hospital and had been admitted without any difficulty.

Afterthoughts: Hospitalization did not end her and her family's troubles. In the short run it may have intensified them. The point had been to provide for her immediate safety, which was accomplished. After her short hospital stay, I saw her on an at-least weekly basis. I think she continued to resent being hospitalized, but I think it helped set the context and tone for our ongoing work that I explained my genuine reactions and decisions; listened with empathy; continued to accept her, if not her intent to kill herself; and left her as much self-responsibility as possible.

AN EXAMPLE WITH A CLIENT WHO IS AT RISK OF HARMING OTHERS

In this third case example for this chapter, we want you to see how the counselor assesses and helps the client plan to avoid the risk of harm from the client to another person, while maintaining critically important therapeutic relationship skills, in hopes of both understanding the truth of the current danger and laying the foundations for ongoing work. The following story is again told in first person, even though we have changed the situation from that of any one client to be like that of quite a few clients. Telling it in first person allows us to write it with the emotions and thoughts that we have experienced in such situations. The names are fictitious.

> John came to counseling at the insistence of his girlfriend, Denise. He explained that she had insisted he come because they had both been drunk at a recent party and he expressed strong anger toward her. He was verbally abusive, yelling and calling her very harsh names. He also threatened her physically, and she had been understandably scared that he would hurt her. She insisted that he get help in order to avoid such behavior in the future. At first, as he explained these things, he seemed to work hard to maintain a casual air about his dilemma. Picking up early in our conversation, our dialog went something like the following:

Counselor: So, you seem to be saying two things at once. You remain casual when you tell me how you acted, as if it was no big deal, but you also seem shocked by your actions.

Client: Well, yeah. I sure didn't mean to be that way, I was shocked. And I know it was a big deal. Everyone stared. Denise might not see me anymore. I was totally stupid, but it was also the alcohol [taking a momentarily dismissive tone, when mentioning the alcohol].

Counselor: So you go back and forth. You know from Denise and others' reactions that it was a big deal, you know it was stupid, but then again, you know it was partly the alcohol.

Client: Yeah, alcohol makes people crazy sometimes.

Counselor: So, you figure it made you crazy that night.

Client: Yeah, well, like I said, not just that. [After a momentary pause] See, three times she was talking to her old boyfriend. He's always boastin' how he could have her back anytime. I just got so jealous. She should know not to do that.

Counselor: So, your jealous feelings were intense.

Client: Yeah. I mean that's no excuse. God, I can't lose her. But I can't stand that bastard. I could just kill him.

Counselor: Such a mix of feelings. You know that you could hardly stand to lose her.

Client: She's the best thing that ever happened to me. She's beautiful, so sexy. And I'm, well I'm all right, but I'm lucky to have her. Any guy would want her.

Counselor: Seems she's so great that while you know you're all right, you see it's almost a dream for you to be with her.

Client: Yeah, like it's not real. Now, it might not be. She told me that if I *ever* do anything like that again, she'll break up with me; she'll never speak to me again. She told me I humiliated her, and she can hardly face her friends now.

Counselor: [I notice about this time that he has dropped the casual way that he tried to tell of his actions early on]. And now you're afraid of losing her. You feel guilty for what she feels.

Client: I feel awful. Man, she's sweet. She's been so patient with me. And the things I called her.

Counselor: So you're really feeling great remorse, telling me and yourself that she didn't deserve the abuse.

Client: Yeah, she didn't deserve it. I'll do better.

Counselor: Again, I get the idea that you have a mix of thoughts. Now you seem to take it all on yourself, but if I remember right, a few minutes ago, you were thinking that she should have known.

Client: I'm not making any excuses. I know I was wrong and that's that.

Counselor: So, you do accept full responsibility but still there's some lingering something.

Client: Well, come on. She was drinking too, and I don't know what she might do when she's drunk.

Counselor: I gather you mean she might have some sort of affair, cheat on you.

Client: Oh, no. I know that's not true. But all the girls love that bastard Jason. God, I hate him. He thinks he can just have anyone any time.

At a point a little later in our session, after taking as much time as possible to know him but leaving enough time to assess and plan for safety, I interrupted and changed the subject in order to let him know my concern and began to assess and plan. Picking up again just before that point:

Counselor: So, you've let me know how guilty you feel, how you mean to take absolute responsibility for your actions, and how, even before last weekend, you've had some uncertainties about your relationship.

Client: Well, she really kinda outrates me, but now I can't live without her.

Counselor: While I very much want us to continue your work in discussing these things, I also want to raise a couple of my concerns of immediate safety with you. [John nods his consent.] I have some concern that you might take action to hurt Jason.

Client: Oh, I'd love to, but I know I won't. Even if we got in a fight, people would break it up. Besides, he wouldn't fight anyway. Why should he? He's gets everything he wants, or almost everything anyway.

Counselor: So you don't actually even have any specific thoughts of hurting him.

Client: No, he'd never fight me by himself anyway. He's always got his friends with him.

Counselor: My other concern is that as guilty as you feel, you might hurt Denise. [John looks taken aback] Well, what I mean is that I notice the passion you feel towards her, how much you want her, and I worry that could quickly turn to jealousy and anger again. [after a pause] You seem okay that I'm saying these things.

Client: Look, I've just gotta not screw this up. I'll do anything to not screw it up.

Counselor: So clearly you want to make sure [John nods yes]. Will you let me ask some questions that occur to me to think about toward making sure you don't do something else abusive toward Denise?

Client: Yeah, okay. Ask away.

Counselor: I know that alcohol reduces a person's impulse control. As you said, it makes people crazy. I want to know if you can stay away from drinking, at least until we meet again in a week.

Client: Well, I won't drink around Denise, but I already have plans with the guys.

Counselor: I gather those plans would be hard for you to change [he nods his agreement]. My concern there is that you or she might find each other, after the time with the guys, when you've been drinking. Do you agree that's possible or likely?

Client: Yeah, okay.

Counselor: Now, I imagine that that part might be hard, to cancel or change your plans with the guys this week.

Client: [speaking almost inaudibly] I already look like an idiot.

Counselor: [responding softly] That part is hard for you. It feels lousy.

Client: [lifting his tone, with determination] Yeah, but it's worth it for Denise. She is *so* cool.

Counselor: That part keeps you going, gives you strength for this work [John smiles his agreement].

Client: My best friend Jarred will understand. I'm gonna tell him the whole truth. He saw it anyway. But then I think I'll have him and me tell everybody else that I'm just staying in—a "getting serious about school" phase.

Counselor: You sound definite about this plan. I have a couple more questions I'd like you to think through [John nods his readiness]. Has anything like this ever happened before? Have their been times when you've lost your temper and become physically threatening, abusive, with Denise or anyone else?

Client: No, never. [pauses] Well, I've wanted to. I can't stand the thought of her with anybody else.

Counselor: So, you've not gotten so mad or acted that way, but you've hurt enough just thinking of her with anyone else that it was kinda close.

Client: Yeah, but I didn't do anything.

Counselor: So, you've never acted that way before with Denise. What about with others in your life?

Client: Well, I've never really had a girlfriend before. [after a fairly long pause, several seconds, in which it seemed he was going to say something more] But me and my mom used to really fight.

Counselor: That was hard for you to say. And it seemed important to you.

Client: [pausing first and speaking softly] Well, I guess I'm ashamed of that.

Counselor: That's very hard for you to tell me about. I'm guessing that maybe your behavior was similar then, since it came to mind when I asked.

Client: I never threatened her. I just screamed and felt that mad.

Counselor: [knowing this from John's way of telling this part] And it still feels lousy to you. John, in the long run you can change the things that give you impulses to act in ways that leave you feeling so lousy. To help in the short run, think of what you feel like just before you get so mad.

Client: [after a long and thoughtful pause] I feel small, like a little kid.

Counselor: You seemed to be able to recall that exact feeling [John nods and shrugs as if to say, "I sure do"]. So, if you get that feeling around Denise, your mom, or anyone else you may have very strong feelings toward this week, then I want you to either remove yourself from the situation, if possible, or at least stop to think and realize that you are moving into a dangerous moment, that you might lose your temper and act in hurtful ways.

Client: Okay.

Counselor: Now I worry that it may sound condescending, but I hope you'll let me review what we've planned [John nods that this is okay]. Until we meet next week, you won't drink. You'll tell Jarred what's really going on, then he and you will tell the guys that it's school. Then, if you get that feeling of feeling small, like a little kid, you'll know that might signal upcoming rage, and you'll either remove yourself from the situation or stop and think, be aware and watch that you stay under control.

Client: I can do this. I can't lose Denise. Thanks.

There are several things we'd like you to note from this example. From what we've told so far; it would be hard to know what would happen next.

I believed his sincerity and noticed that he seemed to get dramatically more serious and real as our session continued and therapeutic relationship developed. I also thought his plan was specific, workable, and voluntary. He left that meeting with a plan to stay away from a trigger for violence (alcohol), and an idea of what it feels like just before he becomes enraged, so that he has a better chance to avoid the rage. I accepted as truth that this really was the first incident close to violence.

Planning for safety with John will be ongoing. In counseling, he seemed ready to begin to realize some of his self-criticisms, shames, thoughts, and feelings behind his anger. Well beyond keeping him safe with Denise, the personal development springing from that work can help him be much more than safe in the future. It can help him become a partner able to love from a position of knowing his self-worth, having faced his shames and achieved a real self-acceptance. The potential imminent danger situation may have brought him to counseling, but the counseling focused in a therapeutic relationship may bring him a much richer and fuller life.

Also, it may be helpful for you to know that it was tempting to me to argue with him early on when he attached blame to things outside of himself. It was also tempting to argue and try to prove how serious the situation was, when he initially seemed to speak of it casually. However, as soon as I realized these impulses, I reminded myself to focus on him as a person and his experiences in that moment. That got me back to responding to him with genuine, deep empathy and UPR. Responding to him in that way helped him break through his understandable defenses and helped tap into his motivation that was there all along to develop to his fullest potential.

Finally, while this chapter has provided three examples of potential imminent danger, there are many, many more possibilities, and we know of no checklists that encompass every possibility. So, your counselor judgment is necessary. Try not to handle such situations alone but to consult with peers. If alone, do your best with the plan to think through the dangerous situation, what makes it dangerous, and how the danger might be avoided, prevented. Then consult. If in consultation or review, you think you missed something important, bring it up at the next meeting or on rare occasions where the thing you decide you missed seems too important, contact your client before the next meeting. Ultimately, it is not possible to do such work perfectly. We believe that your moral, legal, and ethical obligation is to do your best and to take the actions that any other reasonable counselor would. But even when this responsibility is so grave, we want you to see from the examples we've offered that you must never lose sight of the therapeutic relationships you are developing with your clients, that your therapeutic relationships not only help you work through the current crisis but far and well beyond.

ASSESSMENT FACTORS IN DETERMINING RISK AND SAFETY PLANS IN DOMESTIC VIOLENCE SITUATIONS, ESPECIALLY THOSE THAT RISE TO THE LEVEL OF IMMINENT DANGER

In case it is helpful for you to think through the factors we often consider in assessing for safety in domestic violence situations (which also apply to dating violence, as in the last case example), we offer brief thoughts on these factors here.

Physical Violence

While domestic violence situations seem always serious and potentially dangerous, such situations may or may not equal imminent danger. Some may be chronically unhappy, dysfunctional relationships, needing the help that can come from therapeutic relationships but not immediately threatening. In order to help a client discern which might be more true of her relationship, we would want to know if there has already been physical violence. Threats or perceived threats or intimidation are serious, but once actual physical violence has occurred, this seems for us to have crossed a threshold to greater danger. Even if the initial violence was minor, we worry that past violence may make future physical violence more acceptable and more likely. However, we would also want to consider how the couple has reacted to the violence. For example, perhaps the couple reacted to the violence by realizing that their relationship is in great danger and have decided to make consistent and ongoing efforts to change. Or, perhaps they've told each other it would never happen again, without seeming to institute any real change.

The Extent of Physical Violence and Any Potential Pattern

We would also want to know the extent of the physical violence. Actual hitting and previous injuries from violence leave us much more worried for escalating violence than grabbing or pushing. However, it is very difficult to discern this difference and there are no absolutes rules. Through listening with deep empathy and UPR, you can discern how fearful your client is (assuming your client is the one more in danger), even through potential bravado or attempts to make herself and her spouse sound not so bad. Then this understanding of your client's fear can assist you in helping your client assess her safety.

As for the pattern, we would want to know if there seems to be any pattern in the physical violence that might help client and counselor discern the level of danger. For example, is the violence escalating? Is it recent? Is it becoming more frequent? To discern the meaning of such information, we again rely heavily on our therapeutic listening, deep empathy, and UPR to help our client discern the true meaning of the information. We find that very often our clients know their level of danger but only know it deep within their being. So, our therapeutic relationship skills help clients realize and fully admit what they already know, and then resolve to act on this knowledge, sometimes spurred by having discovered a level of self-worth that makes self-care seem worthwhile.

Triggers and High-Risk Behaviors

Along with discerning any pattern, we find it helpful to identify what seems to trigger violence or dangerous situations. This is helpful to know in planning with a client for avoiding dangerous situations in her relationship. For example, there may be increased violence in the relationship when one or both are drinking or using drugs. If your client seems to be more the victim and your client wants to try to stay and work through relationship problems, then she may need to plan to stay sober and plan for a clear means of exit if her potentially violent partner might not stay sober.

Children

We also work carefully to help clients who seem to be in violent relationships understand that if they have children, their children *are* affected. At whatever age, whether the children appear to understand what is going on or not, they do know that something bad and serious is going on. Children *sense what is going on,* even when parents are trying not to show feelings such as strong anger or sadness outwardly. Again, we find that often once we have brought the effects on the children up, then listened with deep and genuine empathy and UPR, our clients who seemed to be victims of domestic violence realized that at some level they already knew and valued this truth.

Planning for Safety

Such plans may vary greatly, depending on whether a client sees the need to at least temporarily remove herself from the relationship. Any plan for safety must of course include the children. Generally, we find that if a client who we see as in potential but not imminent danger decides that she sees herself as safe enough to stay with her partner, and we see no clear reason to disagree, the safety plan should include: anticipating the danger in order to leave before a violent situation has escalated to the point where it is greatly difficult to leave, and knowing ahead of time where to go for safety and how to get there. As with any safety plan we develop with a client, we would work to help our client be as specific as possible and think through potential difficulties with the plan.

COMMON DIFFICULTIES FOR BEGINNING COUNSELORS HELPING CLIENTS MANAGE SITUATIONS THAT ARE OR MAY BE OF IMMINENT DANGER

Situations of imminent danger are among the most difficult for our students who are beginning counselors, just as they are for us. In order to help you do your very best through such situations, we share some of our understanding of what has been most difficult for our students when they were beginning counselors and our thoughts related to these difficulties. While we hope and expect the work of helping clients manage situations of imminent danger gets less difficult for counselors with experience, we expect that the gravity of such situations never lets the work get easy.

The Seriousness of the Situation and the Weight of Decisions—The Danger Itself

Of course, it is the very seriousness of imminent danger situations that leaves many beginning and advanced counselors feeling stressed and worried. The fact that a situation may involve imminent danger means that death or serious injury is possible. This naturally is a source of fear and worry to us as counselors because we know a person's bringing about his own or another's death is one of the few irreversible decisions a person can make.

A simple-sounding thought reminder has helped our supervisees and us through the difficult decisions involved in assisting persons facing imminent danger. It has helped us and our supervisees through the difficult second-guessing that can occur once the client potentially in danger has left the counseling session in any way short of the near-absolute, temporary safety of hospitalization. This thought reminder is that when any counselor has maintained deep empathy, UPR, and genuineness throughout the time of helping manage the crisis, that counselor has used the most powerful and effective tools available to any counselor. The tools of therapeutic relationships are the primary tools we have. The case management tools of assessing and planning for safety are certainly important but secondary. Fortunately, the tools of therapeutic relationships are powerful and effective.

"What if I Panic and Know What to Do, But I Forget?"

We find this a very understandable and common fear of beginning counselors. Fortunately, we have never found it to be true of competent persons. We find that our beginning counselors surprise themselves by rising to the occasion.

The most helpful advice we know for this fear is to work to see that you are well grounded in the core conditions for therapeutic relationships. Developing these ways of being in yourself will not only serve well as tools for you and your clients in times of crisis but developing them in yourself will also develop your personal strength, focus, and clarity of mind. Developing them in yourself can help you develop the dual qualities of being both solid as a person and fluid enough to meet each client where he is in the moment.

A further piece of advice to ready you to avoid panic is to *practice and then practice some more*. Pair up with peers to practice the skills of therapeutic relationships and helping clients manage situations of imminent danger through role plays as often as possible. Seek as many and difficult clients as possible early in your work, while you have maximum supervision and support.

The Pressure of Never Being 100% Sure

In helping clients plan for safety, you will almost never be 100% sure. As one beginning counselor put it, referring to a client who had completed a non–self-harm agreement with her, "All I have is whether I believe or not." What she meant was that all she had after the session was that she believed that her client intended to keep the safety plan.

However, what this beginning counselor also realized is that it is very difficult to mislead, intentionally or otherwise, a counselor who is deeply focused in genuine empathy and UPR. If, through her deep empathy and UPR, she had heard some reservation or, through her self-awareness allowing her to be genuine, something in her client's way of agreeing had left her with nagging doubts, she would have reflected or stated this. So, as she contemplated her work in that session, she realized that while all she had was that she believed her client, her belief was strong, solid, and supported by powerful skills for reaching true trust.

Additionally, while she had thought slowly and carefully in evaluating her client's safety and in evaluating their non–self-harm agreement, she had also empowered her client in these evaluations. She had been open in sharing her thoughts and feelings in evaluating and invited her client into partnership in the evaluations. Thus, her client had participated in rectifying pitfalls in their plans, from the perspective of her expertise in her life and situation.

Discerning the Difference between
Your Feelings and Empathy

Crises and other emotionally provocative circumstances are fertile soil for counselors' reactive versus responsive attention. For beginning counselors especially, it can be difficult to discriminate between one's own feelings, thoughts, and beliefs and those of a client. Differentiating between client and self may be particularly challenging in crisis situations as well as when client concerns strike emotion in you. If you are worried over a client's situation, you may mistakenly assume your client is worried. It is useful to state your concern to your client so that your client knows the level of seriousness in which you see the situation, can respond with her own judgment of the situation, *and* because of saying it aloud and owning it as your feeling, can help you separate it from your perception of her experience. This distinction is important. In some cases, realizing that your client is not worried and why can help you realize that you need not worry so much. In other cases, realizing that your client is not worried can help you see that she is not taking the situation seriously and thus may be putting herself at greater risk.

Errors in Empathy

We have seen two opposite errors in empathy among counselors assisting in situations of grave danger. Some beginning counselors seem to keep themselves blocked from fully realizing client despair, taking action beneath awareness to avoid experiencing client despair. Other beginning counselors fail to see cracks in clients' despair. For example, a client may worry that her boss will be mad *and* also have serious thoughts of suicide. While that bit of possible future orientation does not make her situation much less dangerous, it is important to respond to it. We think of such hints of hopefulness as cracks of light in a dark room. They do not make vision clear or the room bright, but they do make the room contoured and detailed versus solid and dark. We remind you to maintain your empathy with depth and nuance as much as you can through each crisis.

Preoccupation with Liability

A typical preoccupation of some counselors is expressed in the common lingo "CYB" (cover your b___). This mindset perpetuates counselors' focus on their own liabilities rather than on the persons of their clients. We urge you to be reasonably aware of your liability but to maintain your primary focus on the person of each client. Keeping that balance will be both safest and maximally therapeutic through the short and long term of your work with clients.

Having to Let Go and Let Clients Be
Responsible—Trusting That Each Client
Will Actually Do the Plan Agreed To

Beginning counselors have communicated to us their realization of their ultimate need to trust their clients. By this they meant that once they have trusted their client's intention for safety and believe that they, their client, and, whenever possible, a peer or more experienced colleague have developed the reasonably best safety plan, they cannot personally watch over their client's safety between counseling sessions. So, short of taking societal measures that override client self-responsibility (e.g., hospitalization, mental health arrest, warning potential victims), counselors must ultimately trust their clients to be self-responsible. Once that decision to trust clients' intention for safety is established, the ultimate decision is out of the counselors' hands.

Each client's commitment to keep safety plans is based on the strength of the personal connection, the therapeutic relationship established. Fortunately, as a strong personal connection, a strong therapeutic relationship equals a strong client commitment that can see them through a safety plan from one counseling session to the next.

Self-Confidence and Self-Perception
of Competence to Make Such Decisions

Many beginning counselors struggle with having the self-confidence or self-perception of competence to make decisions for client safety. Fortunately, counselors are not charged with making such decisions. Rather, they are charged with helping clients make them.

But even beyond that, you may be able to review the evidence of whether or not you have such competence. For example, is there evidence that your other counseling skills are adequate or beyond? Is there evidence that you are otherwise a competent decision maker? If not, you can work to achieve such competences through your own counseling, through your studies, other personal development, and practice. If you find you do not have the evidence to reasonably conclude, then seek the feedback you need. One way to seek this feedback is extensive practice under observation, and another is initial work under very close and careful supervision. Use your advisors, mentors, loved ones, and your carefully considered intuition to guide your decision of your readiness and your path to proceed.

Coordination with Other Professionals—
Fearing Breaking Trust to Ask for Help

Some beginning counselors have let us know that they were hesitant to consult with other professionals (peers or workplace supervisors) due to concern over breaking a client's personal trust in privacy with them. However, the thought behind this hesitancy is flawed. Each of your clients should be able to trust confidentiality with you. But this is not the same thing as privacy. Suspected imminent danger is a reason to break confidentiality, and this should have been explained to your client initially. Plus, frequently and ideally, the person invited in to consult on the plan for this client's safety would be another mental health practitioner or a supervisor from the same setting, and regular consultation for the improvement of your work can be an additional limit to confidentiality identified when beginning with each client.

There can be an added nuance to fearing a breach of privacy. We and other counselors, both novice and veteran, have experienced workplace situations in which we did not entirely trust the way the peers or supervisors in whom we would be consulting may influence the situation. If such a lack of full trust exists in the long run, it behooves each counselor to try to work through it. Discussing differences and getting to know the peer or supervisor in question better can be one way to work through that lack of trust. Discerning which peers and supervisors seem best to seek in which types of situations can also be important. Then, if this lack of trust can't be worked through or around, seek to make changes in the workplace such that you can have a more comfortable situation for needed consultation. However, in the short run, you cannot let such a lack of trust allow you to isolate yourself. Regardless of whether you fully trust your peer or supervisor, it is best not to finalize decisions for safety over which you are not already quite certain, alone.

Coordination with Client Loved
Ones or Significant Others

Many safety plans require coordination with client loved ones or significant others in clients' lives. In some plans, you and your clients may decide for you to contact persons who will be involved in their plan. Some clients may prefer for you to initiate a contact and request for help due to their own inhibitions for making these contacts. When we've felt nervous and hesitant to do this, perhaps worrying over whether we were ready to add that additional skill or over what the person might think of us, it helped us to remind ourselves that it is surely easier for us than our client to make the contact in that situation. It has also helped us to remind ourselves that our skills in consultation are likely to be more developed than our clients' are, owing to our training in genuine and empathic therapeutic communication. It has also helped us to remember to use our skills of therapeutic listening, empathy, UPR, and genuineness in the consultations. These skills go far in establishing alliances in consultation relationships just as they do in counseling sessions. And finally, it has helped us to jump in and make the contact, without too much overthinking, in the situations where our clients and the one of us involved in the situation have decided it is best.

The Infinity of Unknown or New
Situations for Which There Is No Script

A fear that many beginning counselors share is that they cannot be prepared for every possible scenario of imminent danger. With numerous practices, beginning counselors can prepare for a great many scenarios, but we agree that counselors cannot possibly think through in advance all possible imminent danger situations. Fortunately, the principles of helping in each situation we and many others have encountered have been quite similar. Therefore, it is our belief that it is very helpful to practice for the unexpected but not necessary to have practiced or thought through every possible scenario. In situations that strike you as quite new, it will be helpful to stop and think of what you know to be universal principles for helping manage situations of imminent danger. You *can* and *should* give yourself moments to stop and think.

Shifting into Crisis Management Panic Mode
and Forgetting to Continue to Build and Use
a Therapeutic Relationship with Each Client

This could be understandable as it is difficult to do two sets of things at once. Certainly making such a shift would greatly diminish your effectiveness and heighten the danger for your client. But fortunately, as you practice, your genuineness, empathy, and UPR in counseling can become like breathing—a part of you, something you always do. As you develop strong therapeutic relationship skills, the error of forgetting these skills will be an extremely rare error, only momentary at most.

ACTIVITIES AND RESOURCES
FOR FURTHER STUDY

- Deeply consider, contemplate, and discuss each of the common difficulties for beginning counselors helping clients manage situations of imminent danger that we discussed. Consider which do or may apply to you and why. Create explanations of other difficulties that you expect might affect you and other beginning counselors. Discuss and explain how you may address each of those difficulties.

- Practice making (thinking through and writing out) non-self-harm plans for a wide range of client situations with suicidal thoughts or plans by exchanging possible scenarios with peers.

- Practice making (thinking through and writing out) safety plans for a wide range of client situations that put clients at risk for harming or being harmed by others, including dating or domestic violence situations.

- Journal or essay regarding the need for empathy and UPR in helping clients manage situations of imminent danger. Imagine numerous realistic scenarios taking place in various settings in which counselors may be helping clients manage situations of imminent danger. Explain and illustrate the need for both of these core conditions in each of the various scenarios of managing imminent danger that you have imagined. Broaden your views and answers through discussions with peers.

- Practice using your therapeutic relationship skills in helping clients manage wide varieties of potential imminent danger situations (including mild and severe suicide risk, danger to others, and domestic or dating violence) in role plays with peers. Practice handling these role-play situations without the assistance of a consulting peer. Then, when you have a third partner participating as observer, have that partner play the role of consulting colleague, who comes to consult when you and your client believe you have worked through the situation as far as possible and are ready to consult. Plan with your partners to have role-play practices with clients experiencing mild suicide risk (such that you may practice implementing non-self-harm agreements), experiencing much stronger risk and need for hospitalization, experiencing other types of potential imminent danger situations, and clients bringing unknown situations.

- Try assessing and planning for safety with a role-play client *without* using your therapeutic relationship skills in order for you and your partner in the role of the client to feel how this would be different. Discuss or journal your thoughts and feelings of the difference.

- Research factors described by other authors for assessing risk of imminent danger due to suicidal thoughts and other modes of imminent danger. Compare and contrast these factors with ours.

- Research aspects of non–self-harm agreements and safety plans described by other authors. Compare and contrast these aspects with the aspects that we describe.

- Discuss the issue of hospitalization for mental health reasons in our culture and what you expect it may mean to your clients in imminent danger and to you in assisting those persons.

- Revisit your answers to the Focus Activities for this chapter and write or explain how they have changed based on your studies and practices through this chapter.

- Review the Primary Skill Objectives of this chapter, checking that you have mastered each to your satisfaction at this time. Reread, seek supplemental readings, and repeat practice activities until you have mastered each to your satisfaction.

11

Ending Therapeutic Relationships

Parting is inherent in all meeting. Transience is what gives
life poignancy. Every person is responsible for himself or
herself. There is no road to walk but your own.

TAOIST MEDITATION

PRIMARY SKILL OBJECTIVES

- Understand and accept the principle of independence and be able to explain how it is woven throughout the skills of therapeutic relationships.
- Understand and be able to explain and apply the concept of planfulness
  with endings that are natural and with endings that are arbitrary.
- Understand and be ready to apply the skills of planful endings through
  natural and arbitrary endings.
- Be ready to explain how to and to able to implement the skills of reviewing for client progress, satisfaction, and decisions toward ending within
  your core therapeutic relationship skills.
- Understand, generate examples of, and be ready and able to recognize and
  reflect in review the many varied forms of progress for clients through
  therapeutic relationships.

- Be aware of the pros and cons of alternate modes of planful endings to apply in your practice.
- Understand the problems that arbitrary endings can bring and how to be planful in order to empower your clients through arbitrary endings.
- Understand and be able to explain the context and meanings of the special problems that we present as sometimes occurring around arbitrary endings.
- Understand and be able to explain the context and meanings of common difficulties for beginning counselors around the different kinds of endings that we present.
- Be able to describe commonalities of the great majority of endings of therapeutic relationships.

FOCUS ACTIVITY

If you have been a client in counseling that involved a deeply therapeutic relationship, contemplate what it was like for you to end your sessions and end or change your relationship with your counselor. If you have not yet had such an experience, then imagine what you expect this ending would be like for you. What did you or might you feel related to the ending (a mix of feelings is likely, of course), and why? What was or might be primary and secondary among your thoughts, and why? What are some of the ways that you did or might act, in and out of counseling, around this ending?

Also consider what your reactions have been to ending other close, personal relationships in your life. What did you think, feel, and do related to the ending? How do you expect these reactions may be similar and different from ending therapeutic relationships in counseling?

Now generate and contemplate a wide variety of reactions (feelings, thoughts, and actions) that you expect clients may have related to endings of therapeutic relationships. Then, generate and contemplate the wide variety of reactions that you think you and other counselors may have to ending therapeutic relationships. Journal or essay and discuss with your peers your answers to each of these questions.

INTRODUCTION

We sometimes think of ending therapeutic relationships as difficult and as including painful emotions. But really, that's probably not usually true. In fact, we just like saying hello much more than goodbye. And we dread the goodbyes, at least a little, even when we know they are best and the time is right.

When a therapeutic relationship has been well built and maintained, often the client and counselor know that it is time to end, and doing so can be more joyous and satisfying than difficult. Oftentimes, other uses for the meeting time simply become more motivating to the client than time in the therapeutic relationship. While in the beginning a client's seeming reluctance to attend a session may mean avoidance of hard work, near the end, it can be a very positive sign.

A client's readiness to end can be like a season changing, like winter to spring or summer to fall. We may miss the season that is ending but still know it is time.

It can be like the end of an era in a child's life. While a part of us may want a child to stay innocent and young, autonomy and adulthood remain our goals for the child. The coming of autonomy and adulthood are the ideal of life's development, so it is best to plan for and enjoy the changes along the way.

When a therapeutic relationship has been well built and maintained, ending is not a complex, new skill set for the counselor or client. Rather, it can be simply letting the therapeutic relationship run its natural course. That relationship is always meant as a temporary assistance, helping each client find his footing and continue on his unique path.

The endings of therapeutic relationships in counseling can come in a many different ways. In some situations, after ending individual counseling, you and your clients will have ongoing relationships. Perhaps some of your clients may continue their work and some contact with you in group counseling. It is normal for school counselors to have ongoing relationships with clients after ending intensive individual counseling. As a school counselor, you would usually continue to see former individual clients in the school community and continue to have a role as one of the guiding adults in your clients' lives. In most settings, clients have the option of restarting counseling at a later time, if needed or wanted.

Some endings may be arbitrary, based on the end of an academic year, the client or counselor leaving the setting, the end of available services, or other artificial interruptions. While this is not ideal, it can be planned for, and progress in the therapeutic relationship can be consolidated and continued beyond the regular contact in individual counseling.

Sometimes your clients may get to work through some of the grief of ending their therapeutic relationship with you while still with you. Even when this is not openly discussed, it is often a part of the ending, and you may watch for and address it.

Above all, of course, we wish for you to maintain your therapeutic relationship with each client through to the last moment. We offer guidance in the coming pages on how to do this, as well as how to add skills that are helpful for ending.

THE PRINCIPLE OF INDEPENDENCE

Work for the end from the beginning. Your goal with each client is independence. While powerful and important, you are only a temporary part of your clients' lives.

This overarching principle of independence, of having your clients end their time with you in counseling maximally self-reliant, is why you do not give advice or take control of clients. It is why you work through the process of developing therapeutic relationships, which may well be harder work than attempting to exert control over clients, harder than attempting to do for or live for your clients. It is like watching a child struggle to master a task that you know you could just take from her and do for her and quickly end the child's struggle with the task. In doing so, the task would get quickly done, but the child would have still not learned to do the task for herself and would likely have received a message of your seeing her as incompetent. Controlling and doing for clients would leave your clients always in need of you. We remember a colleague who frequently wondered why he had so much difficulty getting clients to end counseling—it seemed to be that his way in counseling taught dependence, by advising, controlling, and thinking and doing for clients. Working through the process of developing therapeutic relationships helps your clients to continually develop self-responsibility, to end in a state of independence from you and interrelated autonomy with key others in their lives.

While you may love your clients, feel great warmth and caring for them, it should be a nonpossessive love. You should love them every bit as much when they are moving on from you as you did when they seemed to need you greatly. Again, you are only a temporary part of each client's life.

PLANFULNESS

One of your tasks from nearly the beginning in each therapeutic relationship is to determine when the end may be. When external factors will control the ending, such as a limited number of sessions allowed or the end of an academic term, this limitation should be made clear to each client near the beginning of your relationship and should be taken into account as you structure or explain each client's use of time in relationship with you.

If the ending will not be arbitrary and the therapeutic relationship can reach its natural conclusion (which will not necessarily take a long time— some therapeutic relationships reach their natural conclusion quite quickly, perhaps after only one meeting), discern when your client expects that ending to be. You can usually do this when you are explaining how you see that the interactions that make up a therapeutic relationship and its outcomes apply to that person's concerns. Whether a part of this explanation or not, at some point early in your time with each client, discern what you think each client's time expectations are and reflect what you think you see.

For example, if a client has met with you and vented or explained his frustration with a romantic partner, parent, teacher, or others, and you have listened therapeutically and found it easy to meet him with genuine empathy and UPR, and he seems quite satisfied before the end of the meeting, you might reflect and make structuring statements something like, "John, I have the impression that you've gotten what you'd hoped for in our time together, that you wanted this confidential setting to tell your concerns, to sound them out and learn from them. I would want you to tell me if it is not the case, and we could plan to continue, but I'm assuming from your seeming to be satisfied and sure of your next steps that this is the only time you're expecting us to meet, at least right now." That statement is long. So, of course you would allow for and attend to interruptions or apparent strong reactions that your client may have to any parts of it.

Committing to Review for Client Readiness to End throughout Ongoing Work

For an example in which we might expect the relationship to be ongoing, imagine that your client seems quite troubled and dissatisfied by the things she has come to communicate. Within your first meeting, you might reflect and begin to structure the ending, saying something like, "Jennifer, I get the idea that this situation has been troubling you for some time, that it hurts a lot right now. I would be glad for us to continue beyond today and I have the impression that you would, too." Assuming she affirms this, you might then suggest, "Let's plan ongoing weekly meetings and plan to periodically review your progress, if you are getting what you hope for from our time, and your decision of whether and how long we might continue."

Please note that this explanation assumes that, with your input, it can be your client's decision on whether and how long to continue. If that is not the case for some setting-based reason, such as a limit on the number of sessions or the coming of the end of the school year, then include those limitations within your structuring statements. For example, you might amend the statement to say, "Let's plan ongoing weekly meetings, for as long as the 12 weeks until the end of school [alternate example: "for as long as the 12 meetings that we may have"] and to periodically review your progress, whether you are getting what you hope for from our time, and your decision on whether and how long we might continue within those 12 weeks." Again, as with any such long statement, allow interruptions and attend to client responses to any part of it.

For another example, perhaps the things your client has come to communicate are not yet clear to either of you. Such situations are fairly common. In that case, your statement that reflects and begins to structure the ending might sound something like, "Sara, I get the idea that you are motivated for our time together, that there are things you want to communicate, and that we are only just getting started. I'd like us to continue, and, correct me if I'm wrong, but I've had the impression that you'd also like us to continue. So, I'll suggest we plan ongoing weekly meetings and to periodically review your progress, whether you are getting what you hope for from our time, and your decision on whether and how long we might continue."

Reviewing for Client Progress, Satisfaction, and Decisions toward Ending

Now that we have committed to periodic reviews, you might wonder just how these reviews happen. So, we provide explanations and examples.

While we may find that we review progress, satisfaction, and expectations for ending with our adult and adolescent clients about every three meetings (with children, we see this as an adult decision, with the child's way of being/behavior providing input), we don't believe a definite amount of time should be prescribed for this. Rather, such reviews can be made in reflection when they occur in the flow of sessions. If you have it in mind to look for opportunities to make such reviews, then when your client communicates things that indicate progress or that indicate his reaction in review, you can see them and respond.

Reviewing for Progress For example, if your client tells you of taking actions outside of counseling that you know to be within a direction of change that she had hoped for, you can empathize with her experience and reflect something like, "[with a smile and relieved tone that matches the client's in that moment] I remember that you'd feared and hoped to reach out to others and now that you've taken steps to this, you seem very pleased with yourself." Please realize with this statement that first steps may not mean it's time to end. Normally, paths to progress are filled with new steps, missteps, and falters. Alternatively, you also may empathize and reflect lack of progress, "You still feel depressed and you are aggravated with it, frustrated that it's ongoing!"

Reviewing for Satisfaction Review regarding whether your client is getting what she hoped for can come as you notice your client's process in her use of counseling. An example reflection might be, "Sara, I get the idea that you've been working hard, risking looking at parts of yourself that seemed scary to you, and while this is hard, you are motivated to do it."

Alternatively, you might reflect a lack of meeting client hopes, "Sara, I see that you are frustrated with our work. You had hoped I could give you answers that would work quickly, and you are finding that's not the case." Both empathy with encouraged and discouraged feelings are examples of counseling that is working. They are a part of the work with that particular client. Clients need to hear their frustration expressed aloud to fully realize it. A client may also need to understand her frustration and her counselor's inability to provide quick easy answers as part of her choice to more fully dedicate herself to the very hard work of finding the answers that are unique to her and that she truly owns.

Reviewing for Client Intentions toward Ending Regarding reviewing client expectations for or readiness to end, an example of a reflection and structuring statement that reviews this is, "I notice that you seem to have gotten a lot of what you hoped from our time and that it's getting harder for you to schedule the time, as other things are becoming more important. I'm going to suggest that we begin to end our work together." Then after listening to your client's response, you may suggest a time frame based on your perception of his intentions and readiness. Your client may have responded to your reflection and time suggestion by saying, "[with a brief, sheepish expression] Yeah, our time has very important to me. I, I'm going to miss it. But, yeah, these other things are also very important to me now. I just don't quite want to stop our time yet." So, you might respond, "We seem not quite at an ending time, but very close, so I suggest we plan for four more weekly meetings." Or, if the events that seem to be becoming more important than counseling sessions and your client's attitude suggest it, you might offer something like, "I have the idea that you value and enjoy our time, that you seem to see yourself as ready but not wanting to end. So, I'd like to suggest that we plan to meet three more times but that we meet every other week."

With clients who are not near being ready to end, reviewing intentions toward ending usually takes the form of saying nothing at all and assuming the client's intention to continue from her way of being. The review may take the form of your reflecting and structuring by stating your intention or desire to continue. An example that combines a process reflection, a statement of the counselor's view of counseling for that client, and an understanding of that client's intentions toward continuing is, "Sara, I see that this work is hard for you, and that it hasn't yet produced what you want. But I remain confident that our work can benefit you, and I want us to continue this work."

Setting a Tentative Plan for Counseling, Reviewing Progress, Satisfaction, and Decisions about Whether and How Long to Continue with an Initially Reluctant Client

You may also serve clients in whom you or others in their life see the need for them to work in counseling, even when that person does not. For example, there may be situations in which a spouse, parent, or teacher wants a person to use counseling with you, but the person who is to be the client has misgivings. In such situations, it is of course imperative to listen therapeutically and meet your client with genuine empathy and UPR, but in addition to that we have found it useful to make structuring statements similar to those just given in order to set a time frame for continuing. An example of beginning to plan for the ending might flow as follows (The example begins in the reflecting process and continues with an example explanation of how counseling might apply to a reluctant client):

[Picking up in the last minutes of the initial session.]

Counselor: So, while it wasn't fully your choice to come and meet with me, you seem like you might be interested to know just what possible benefits our time could have.

Client: Yeah, no offense, but I just don't see the point. Yeah, I get mad sometimes, but this is just so unfair. I am really tired of it. I've just about had enough!

Counselor: You're aggravated and mad about it now!

Client: Yeah, so enough already. Tell me, what's the point in continuing? I'm busy and I could just go be mad somewhere else.

Counselor: Well, at this point, I have a few thoughts, but I don't know for certain. You find yourself in situations that make you mad but that you don't see any way of changing. A short-term use for counseling might be for you to express some of your anger to keep from blowing up, like letting some pressure off before too much builds up.

Client: Yeah, maybe.

Counselor: But really, I'd like to work beyond that. For many people, anger is sometimes an emotion that is on top of lots of other things. As we shine a light on what you think and feel in our work together in counseling, you may come to see more clearly the meaning of the situations of your life to you and begin to see more possibilities for yourself.

Client: Yeah well, I see that you believe that, but I'm not sure what the heck that even means.

Counselor: Right now, it sounds pretty strange to you and you're not sure if you are interested.

Client: Nah, it's interesting. You're a trip, but I'm busy.

Counselor: Then I'll suggest this: Let's plan to meet weekly for the next three weeks. That may give you enough time to get an idea of what I'm talking about. Before the end of that three weeks, we'll consider whether you're any more or less interested, and see what you are thinking at that time about whether to continue and maybe for how long. [Slight pause.] As I was talking, I thought your facial expressions were both interested and reluctant.

Client: Well, yeah. I can do the three meetings, but then, we'll see.

Considering this client's openly stated reluctance in response to his counselor's expressions of empathy, opportunities to review his progress, satisfaction, and decisions toward ending may very naturally occur. With such a client who is initially reluctant, we would, of course, remember to meet him with empathy and accept his decision, whether that decision is to continue beyond the three sessions or not. For such a client, it would also be important to remember that while it is within each counselor's role to help most adult or adolescent clients understand their potential use of counseling, it is not within any counselor's role to convince a client to use a therapeutic relationship in counseling. Trying too hard to convince a client would shift the counselor away from providing the core conditions and lead to the temptation to hold out false promises.

Recognizing the Many Forms of Progress

The American Heritage Dictionary (1982) includes these phrases in defining *change:* "to be different" and "a transition from one state, condition, or phase to another: *the change of seasons*" (p. 258). Webster's Dictionary (Grove, 1976) includes these: "to give a different position, status, course or direction to . . . a shift from some mode of personal action or disposition or matter of concern to a different one" (pp. 373–74). The American Heritage Dictionary (1982) includes these phrases in the definition of *progress:* "movement toward a goal," "development; unfolding," "steady improvement," and "to advance to a more desirable form" (p. 990).

So, your clients' change and progress can take many forms. It might be changes in behavior; it might be changes in your clients' ways of being; it might be a reawakening of maturation processes; it might be different ways of seeing concerns or self; it might be changes to different sets of concerns; it might be goal directed, or the steady unfolding or blossoming of your clients' persons. Change in counseling can be both initially inconspicuous and life altering.

Examples of the Many Forms of Progress Please consider the examples of forms of client progress that we offer here. You may be able to recognize progress as an attitude change. For example, as your client comes to trust you and others more, she may look more directly at you when speaking and express emotions more fully and openly to you. A client who had been in frequent fights might let you see his initial progress by expressing his true remorse over having hurt others. Changes may appear more outside of sessions. A client who had been quite fearful may begin to take reasonable risks. A client who had been frequently angry may begin to exhibit a more calm, accepting way both in and out of sessions. Fuller examples of progress follow:

- One client began counseling already realizing that she felt a great anxiety and a desire to conquer it. Early change for her came in reframing how she perceived this anxiety. As she talked about her anxiety and experienced her counselor's understanding and acceptance of her, she realized she saw her anxiety as a huge entity. She realized that she believed that others had tools that she did not for battling anxiety. The more she discussed her thoughts and feelings related to the anxiety she felt, in the light of her counselor's acceptance, empathy, and warmth, the smaller her anxiety came to seem. Soon, her counselor noticed that she developed a way of acknowledging her anxiety when she felt it, then waiting to allow it to pass. As she came to accept her anxiety, she came to accept herself. Her progress became measurable in behavioral terms but was first seen by her counselor as a change in perception and accompanying way of being.

- Another client came to his counselor having ended abruptly with two previous counselors. Initially he seemed to ask and long for a name for what was wrong with him and a quick solution. When his counselor met him with empathy and UPR, as well as suggested how counseling might work for him and her hope that he would engage in a self-exploration process, he was initially taken aback. Initial progress for him was becoming able to accept this notion of help that was very different from what he had expected. A later measure of progress was his realizing and explaining that while he had asked previous counselors to name and solve his problem, he resented the labels and resisted the suggested fixes that seemed out of his control, such as recommending that he consider medications aimed at changing his emotions and behaviors. Well before he articulated these new understandings, his behavioral improvement was under way.

- One client came to realize the extent and hurtfulness of his actions in anger and his dissatisfaction with this, plus the parallel of his anger to his father's anger. He developed a greater awareness of his pain that resulted from his father's actions in anger, its relationship to his anger, his fear of passing this pain and anger onto his children, and his great remorse over that possibility. He realized that his anger seemed to cover up more core emotions that he had not realized but still had feared greatly. From these realizations, he began to dare to express his more scary feelings in sessions. When angry, he began to stop to question what he was doing and feeling and then decide what to express and how so. He found the motivation to begin his path toward becoming the person that he really wanted to be.

- Another client, who let his counselor know that he had been repeatedly sexually molested as a child by a family member, was experiencing difficulty functioning in school and in family relationships, and was suffering from frequent insomnia or nightmares. Through his therapeutic relationship with his counselor, these easily measurable trouble areas outside of counseling improved. But even before clear changes outside of counseling, there were more subtle changes in his ways of being. In his first sessions, there were long silences (almost half the session time), in which his counselor would accept and feel with him in silence, and occasionally express her empathy or understanding of what he seemed to be feeling or seemed to gesture. He also spoke very quietly and softly during these first sessions. Soon, he built trust and came to talk more directly to her. He came to spend more time on subjects that his counselor saw as hard (his current alcohol use, his molestation) versus softer subjects (school). His stories became increasingly more intimate and were told with clear annunciation, animation, and direct eye contact. He came to have a full understanding of what he felt and why (i.e., "I flinch when he tries to hug me." "I feel afraid." "I'm still deciding who I can trust and working to let go, when I know that I really can trust a loved one"). His counselor was able to review many of these changes in reflection. These reviews in reflection (i.e., "I notice your growing confidence in speaking to me." "You understand what goes on with you in such moments and how you want to change") also helped this client track his progress and seemed to help him feel encouraged to continue.

- Another client began her sessions seeming depressed and with her communication emotionally flat, even while telling things that seemed quite hurtful to her. She soon realized and explained that she was dissatisfied that she did not stick up for herself and came to realize that she put undue value in what others might think of her. She came to realize she didn't trust herself or really trust others, even while she was heavily weighting the critical views she assumed from others. She spent much of her early time in counseling telling her experiences of others hurting her. Soon she began to trust her counselor and so expressed more of herself. Her counselor could see her self-trust in the fact that she expressed more of herself, but also in the way she expressed herself. When she spoke, even of troubling, hurtful things, she came to seem full of herself, as if she had been physically insubstantial but then became solid. She began to make direct eye contact when communicating with her counselor. Her speech became full with a range of emotions, rather than flat. She began to take responsibility for her decisions, to act and move in the directions of her goals, even when she suspected that others might well not approve of them. Instead of using her time in counseling to tell of the ways others had hurt her, she gravitated to using her time in counseling to tell more purely of her experiences and discussing her choices and decisions.

- Another client was quite nervous just getting started. As her counselor accepted and expressed her empathy of this nervousness with her, and helped her understand counseling and how it might work for her, the first progress the counselor saw was the client becoming comfortable with using counseling; becoming able to speak more smoothly, rather than fearfully; knowing more what she wanted to say or being willing to improvise freely instead of worrying so much over whether what she said was right or okay. As she progressed, the client came to see herself as holding back too much from expressing herself, both in and out of counseling. Her change began through experimentation with expressing herself in counseling, learning just what she felt and thought about her new level of expression, then carefully moving to express herself more fully out of counseling as well.

- For a client who began counseling expressing belligerence toward counseling and disgruntlement toward even being there, his counselor was able to observe his first progress as settling into his work in counseling and seeming to accept his situation of that moment. Little by little, then all at once, he seemed to come to embrace his work in counseling, to look forward to his meetings with his counselor, and to want to get the most out of every meeting. As this initial progress was taking hold, he began to discuss parts of himself that seemed to bring troubling emotions for him and to change those parts, even without taking the step of describing any parts of himself as flawed, as some might have assumed necessary.

Further Considerations Regarding the Examples We Have Offered
We do not mean to say that all clients' progress, across all settings and situations, will look like the examples we have offered. Rather, we offered those examples to spur your mind to think broadly and specifically for opportunities to see client progress.

It is important to note that with those examples, the realization parts were never analytically interpreted for clients and were often not stated by clients. Although it might be nice, it is not necessary for a person to say such things aloud or even to have the ability to articulate such thoughts. Clients can experience the internal change and act on it, and counselors can see the internal change, even before outward behavior change is clear, as long as the counselors are listening therapeutically, attuned with deep, genuine empathy and UPR.

Also in those examples, counselors tended to reflect aspects of progress in terms of noticing clients' changes in ways of being, as much or more often than reflecting specific behaviors. For example, reflecting a change in eye contact can leave some persons feeling overly self-conscious of their physical self. An alternative is to reflect the different feeling that you feel with your client in connection with the change in eye contact (e.g., "You seem to have become much more confident in speaking over our last couple of meetings").

In a final note, the counselors also used their connections with their clients to help them to know when to reflect in review of progress and when not. Of course, when a client is in the midst of feeling and expressing very strong emotion, that is the time to respond to that emotion, not to the progress that emotion may also represent.

Consideration of Alternative Modes of Planful Endings

Standard Blocks of Time We have known some counselors to frequently plan for a set number of sessions, such as six. We believe the thinking behind that method is that clients will have a greater awareness of the finite and be more motivated to work quickly.

However, our view is that while counseling based in a therapeutic relationship is efficient and can be very short term, planning for specific short segments of time can leave a client expecting some quick solution and feeling wary of letting his counselor know his real self, for fear that they would just get started, then would have to end. On the other hand, if a client began to work in earnest, letting his counselor know his real self, a lot could be accomplished in just six sessions.

Having Clients Decide to Continue Each Week When working with adolescents or adults in settings that facilitated working this way, we have liked having clients schedule for the following week at each session. This seems to have an effect of making clients recommit to their work upon ending each meeting. In such settings and situations, we have tended to end each session by addressing the issue of continuing or not. For example, unless it is just incredibly obvious and therefore redundant or predecided, we frequently end sessions with a reflection and structuring statements like, "I would like us to continue and this seems your intention, too. Come and let's schedule for next week." If we thought a client's intention to continue was less clear, we'd modify the statement and allow more time for her reply, saying, "I'd like us to continue, but I have the feeling you are ambivalent about it."

A problem with bringing up such a decision near the end of the session is that it can take more than a moment or two to decide, and taking more than that moment or two could either interfere with your client's full session and ability to work under his own direction up to the end, or possibly make you late for your next meeting. Therefore, while we like the practice of rescheduling each week, we also like the practicality of scheduling weeks in advance, and reviewing progress, satisfaction, and client decisions related to ending as these topics come up during sessions.

Counting Down to the Ending

Once you have discerned when the ending will be, you have given yourself the opportunity to count down to it. By this, we mean that when there are about five meetings left for you and a particular client, you may begin a session by saying, "After today, we have about four more meetings." Then, in the next session, "After today, we have about three more meetings," and so on. You should make your countdown statements with a tone that indicates that the time and relationship are important to you but that is also neutral. By neutral, we mean you are not indicating with your tone that five sessions is either a very long or a very short time. Some clients will react to hearing that their remaining time with you can be given a finite number and begin to speed up, learning and communicating as much as possible in the time they have. However, your purpose in counting down the time is not to hurry your clients but to have your clients be aware of the time, make whatever meaning of it they will, and then have the opportunity to plan and own full use of their remaining time.

For some clients, this countdown to ending and awareness of remaining time allows them to grieve the loss of their time with you while they are still with you in counseling. However, most often, especially with naturally occurring endings, clients simply hear and acknowledge your marking of time, then move on with communicating experiences of the moment.

Letting Your Clients Know They May Return

If it is possible for your clients to resume counseling with you after initially ending, then, of course, let them know that. For example, when discussing ending or counting down, you might add a statement like, "Of course, if you decide there is more work you want to do in counseling or you want to meet for some other reason, then please make an appointment and I'll be glad to see you."

Telling Your Clients How You See
Them in the End or Last Meeting

Some counselors think of telling clients how they see them and feel toward them in final sessions, kind of like telling a friend how much they mean to you in parting. However, in order for each client to own her use of time in counseling up to the end, it is important that your actions in the last session be quite similar to your actions up to that point. So, there would be nothing wrong with your telling a client how you see and feel in response to her, if you have done that throughout.

Of course, we hope that you have done that throughout, as such statements can be high-level reflections. Helping a client see himself better through your view of him is one of the core mechanisms producing client change in therapeutic relationships. Telling a client how you feel in reaction to him can also be a reflection and a tool for that client to learn more about himself through your reaction. However, this is a complex, very high-level skill. A delicate aspect of this skill is discerning if what you feel is more due to your cognitive constructs, belief systems, and transferences, or if what you feel is more purely a response to your client. For example, do you have negative feelings toward certain actions of your client because you believe that all persons should act in certain ways? Do you feel affection for your client because his respect for you leaves you feeling elated and seeing yourself as effective?

Also, the following question may help you think through sharing your feelings in reaction to clients with them. If you would tell clients of whom you have very warm regard those feelings in response to them, would you also tell clients for whom you are experiencing more negative-seeming feelings in response to them? Of course, it is easier to express the warm feelings, but if you would only express the warm feelings, you must consider whether you are using your expression of your feelings to help your client understand herself or to push some other, less therapeutic purpose, such as trying to convince her of her worthiness or that she should cheer up. Due to the delicate and subtle nature of the high-level reflections needed for this skill, we have included multiple activities for self-development and understanding that relate just to this section with the activities at the end of this chapter.

ARBITRARY ENDINGS

Arbitrary endings are common. By arbitrary, we mean that something unrelated to your client's decisions about ending and unrelated to your client's progress or readiness to end counseling acts to end that client's time with you. This includes such things as the end of a school year, an agency policy on numbers of sessions allowed, or the relocation of client or counselor.

Arbitrary endings require special planning and consideration on the part of the counselor and client, but they certainly do not mean that significant progress has not been made or would be negated by the ending. Years ago, we were at a presentation by a counselor who worked with highly behaviorally and emotionally troubled children (unfortunately, we don't remember the person's name in order to give credit), and the presenter was asked to address the problem of the children he served frequently moving and thus having counseling services disrupted. He, of course, first addressed efforts to contact counselors in each child's new community in hopes of more smoothly continuing services, if possible. He then went on to offer a metaphor that has long stuck with us. The metaphor asks that you imagine that you are traveling in the desert and you cross paths with a person who is thirsting to death. While you may not be able to give this person enough water to supply the rest of his journey, he will surely appreciate the refreshment of what you can give.

Additionally, we know that even a small experience with a therapeutic relationship can have powerful, lasting, and catalytic effects. For example, consider the mindset we see as common to children whose behavior may be described as conduct disordered. We have seen that such children often believe that they cannot be loved, liked, and accepted, and that everyone will reject them. Yet, if you meet such a child with deep, genuine empathy and UPR, especially while allowing that child to let you know who he really is, to let you know real feelings, whether those feelings are easy for you to experience with him or not, then that person's absolute belief system must begin to allow for exceptions and so it may begin to crumble. Once one significant exception to an absolute belief is experienced, the absolute is in great danger of faltering. We have also seen that behind that belief system, another set of beliefs implies that he is inept for and undeserving of human connection (Cochran & Cochran, 1999). Yet, as you invite and anticipate such a connection with your skills, that second absolute belief system must also begin to crumble.

Take for another example a client who feels great anxiety. Imagine that a young woman has seen others for help with this anxiety before. Each time, she asked for suggestions of what to do to make her anxiety go away. Then, the suggestions she tried only worked a little, if at all. It seems likely that she is acting from a belief system saying that there must be an external solution to her distress and that she is not competent to master it without some sort of external solution. It may well be that this belief in her incompetence is an important part of what drives her anxiety.

Then she meets with you, and you meet her with your deep faith in the process of building therapeutic relationships, including your faith in her self-actualizing tendency and the power of her self-discovery to reinvigorate her self-actualizing tendency. In order for her to understand what you offer, you would, of course, explain such things to her in terms that would seem to make sense to her and that are applied to what you know so far of her and what she wants. But most importantly, instead of offering more advice, you offer your trust of her use of a therapeutic relationship. Then, even if the unlikely worst-case scenario happens (this has been very unusual in our or our students' and friends' experience) and she leaves an initial meeting with you thinking you a little odd, she is forced to reconcile her experience of you with her belief system. Moreover, if you only have one or two meetings with her, due to some arbitrary, external limiting factor, while you may inform her of resources for her continued work, the most powerful and lasting thing you can do is to have met her, if only briefly, with your self focused in deep, genuine empathy and UPR, and in building and using a therapeutic relationship in the time you have.

From even a brief interaction with you, if you have made your time worthwhile in deep therapeutic relationship, persons can be forced to reexamine sets of beliefs about others and self that may have driven much of their misbehavior or exacerbated much of their pain. From a brief encounter, there will be no guarantee of change, but at least you know that it will be difficult for such clients to maintain absolute beliefs and the behaviors and emotions driven by those beliefs.

While the results may be more subtle with less extreme examples, the same principle is true for all clients: a brief encounter with you in a therapeutic relationship can be quite powerful in producing ongoing effects. So, while arbitrary endings require additional thought and planning, we, of course, encourage you to engage fully and make the most of the time you have.

Help Clients Plan for the Premature Ending

If you think of your time with a client as limited to less than what you expect she might need or want, make your client aware of the time you have with her and why. Then, as with any structuring comment, be ready to respond to your client's reaction to the information with empathy. Then respond with further structuring explanations, if it seems necessary. An example follows:

[Picking up early in an initial session]

Counselor: Nina, before we get too well started, I want to be sure you realize that there are three weeks left in the semester and that I won't be here next semester. So, we can have about three weekly meetings to work together.

Client: [Seeming hesitant, taken aback, speaking haltingly.] Oh, well, okay, I guess.

Counselor: [Not yet sure whether the client is bothered by this information or just doesn't know what the point is.] You had some reaction to what I said. You didn't seem to know how to respond.

Client: [Still hesitant and speaking haltingly.] Well, no. I don't know what you mean by that. Is that too little time to work? I, I don't know. Maybe I should have started earlier.

Counselor: You're worried that I mean it's not enough time to be useful. No, while I wish we had more time, we can get a lot done in the time we have. I just don't want the end to sneak up on you. So, I'm letting you know now.

When agencies have limits on numbers of sessions available to clients, then this information may be conveyed initially in writing, often within paperwork that clients sign to acknowledge that they have read and understood the conditions of counseling at that agency. In our experience, when that time limitation seemed to encompass the time usually needed for counseling to come to a natural conclusion, we have not seen a need to reiterate that arbitrary ending. Exceptions might be for a client who seems to be working in such a way that you suspect he is not aware of the time limit or for clients who are nearing the end (within four or five sessions) of the time limitation.

Whatever the situation of the limit, you should keep your tone caring but neutral when informing your client of the limit. For example, in reminding the client whom you thought was not cognizant of the time limit, your purpose is not to convey a message to hurry up. Rather, your purpose is to make him aware so that he can make an informed decision of how to proceed. If you have a sense that this client is stalling or hesitating to communicate something that he seems to want to communicate, then that is a different matter and may more honestly and effectively be handled with a process reflection (i.e., "I get the idea that there is something that you came here to say or do, but for some reason you are putting it off, filling our time with other things") and perhaps an explanation of how you see that he could best use counseling (see Chapter 8 for examples).

Counting Down

When the end is coming soon and quite possibly in the middle of a client's work, you should be especially careful to count down sessions toward arbitrary endings. This is to keep clients mindful of the time remaining so that the end is not needlessly abrupt.

Discussing/Suggesting Continued Work and Progress

With naturally occurring endings, it is usually unnecessary to initiate a discussion of ongoing self-development. If needed, the discussion occurs naturally and is client driven. However, with arbitrary endings, we think it can be reasonable and helpful to offer your thoughts on ongoing work. For example, you might suggest that a client consider continuing with a counselor that you recommend, explaining how and why so. Or, if your think a client has begun a particularly useful area of self-discovery, you could suggest that she continue that work in counseling, journaling, or contemplation. Example counselor statements of such thoughts follow:

Counselor: Nina, I see you as having gotten a big start on work that is quite important to you. I also have the impression that you've come to value our work. So, I'd like you to consider continuing counseling with my colleague Susan next year. While it would be something like starting over for you, you'd be starting at a very different place than when we first met. Also, I know Susan pretty well and believe you can expect her work to be much like mine.

Counselor: Nina, I have the impression that the work you've started with me is quite important to you. You seem particularly excited to have begun learning some of the thoughts that seem to drive the anxiety you feel, and you seem to have begun evaluating and even changing some of those thoughts. Perhaps you already know this, but if not, I'd like you to consider that you might be able to continue this work outside of counseling as well. I'm suggesting that you consider journaling your feelings, thoughts that seem related to those feelings, and your thoughts about those thoughts and feelings.

Please realize that these counselor statements are long. So, in a real session, the counselor would need to allow for client interruptions to comment and also would need to stop to attend to client reactions when those reactions appear strong or obvious from a feeling conveyed, a facial expression, a body contortion, or other modes of expression.

Also realize that there is nothing generic about these suggestions. Rather, they are only examples. There could be many more possibilities. The second one would surely require more explanation and a client who seems able to understand and implement cognitive journaling (Beck & Weishaar, 1989) without support from a counselor.

Also, it may be important for us to note to you that we do not make referrals to counselors whom we do not know. If we are only making a general suggestion that a client continue in counseling, we might say something like, "As you seem to be considering continuing in counseling, I hope you can find a counselor that you respect and can value working with in _____ City [where the client is moving to]." Then, it might even be reasonable to help your client know how he might recognize a good counselor for him: "As you seem concerned about whether it can work the same way for you with a new counselor, consider in your first meeting with that counselor whether you believe you are respected, listened to, and understood. If not, I encourage you to say so and then consider the counselor's response to your comment.."

And finally, we want to reiterate that you *may* make such suggestions, meaning you do not have to. We only make such suggestions for ongoing work when the situation seems right and the suggestion quite helpful or necessary. We encourage you to use your judgment in each unique situation.

Special Problems or Situations
That Occur with Arbitrary Endings

Special problems that are much more rare around naturally occurring endings can occur around arbitrary endings. As long as you keep yourself focused in a therapeutic relationship with each client, which prompts and honors client self-responsibility, then most problems with endings are not really problems but are situations for you to consider to help you further think through the planfulness of your work as well as potential client behavior. Examples follow.

Some Clients May Adjust Their Pace According to Their Awareness of the Upcoming Ending Some clients may speed up their work and others may slow down with the awareness that their time with you is nearing an end. By speeding up, we mean that when nearing a planned-for, arbitrary ending, some clients seem to realize that they have much that they want to do in the time they have and so increase their rate of disclosures and self-discovery. Other clients may have the opposite reaction to realizing that their time with you is nearing an end and seem to pull back or begin to detach, appearing to prepare to separate from and to be without their work with you in counseling. Still other clients, and perhaps the majority, may remain mostly steady in their work in their therapeutic relationship with you, right up to the planned-for, arbitrary ending.

It is important to realize that none of these client reactions to an impending ending are wrong, better, or worse. Each person's reaction to the ending of her therapeutic relationship with you is simply that person's reaction, her actions based on the thoughts and feelings that she experiences. While we may like it if our clients speed up their rate of disclosures and self-discovery near the end, this could potentially leave a client feeling open and exposed in the middle of her work when the ending does come. But we respect this as each client's decision that she has a right and responsibility to make. While we may be disappointed to see another client pull back, that person may realize that he needs to detach and reimplement psychological self-defenses in order to weather the arbitrary end of his therapeutic relationship. We would also acknowledge and respect that decision.

You should accept each client's reactions to ending with you, just as you have accepted each client's decisions throughout counseling. You should strive to understand each client's reactions to ending through deep empathy. You should say what it is you see through this empathy so that your clients can be fully aware of their choices around ending and grow from this, as they have from other aspects of themselves that you experienced and communicated. As it is not clear that there is a best way for a client to utilize such an ending, we would not usually explain our views on how a client could best use the time to ending.

For example, with a client who speeds up, you might reflect, "I notice that the closer we get to the end of our time, the harder and faster you seem to work. Seems like you've decided to get the very most from all of the moments we have left."

An interchange with a client who is pulling back might go like this:

Counselor: [Noticing that her client seems to have become less talkative and that there are more long pauses in which she seems to be thinking of things to say and discarding them.] I have the thought that you've begun to pull back from letting me know you. I get the idea when you pause for a moment that you are thinking of things to say, then deciding not to.

Client: [Another pause] Yeah, well I just don't know what to say.

Counselor: So, while you don't know what to say, there seems more to it than that. You seemed to also acknowledge my idea that you are considering things you have an urge to say and discarding them.

Client: Well, I just have these little ideas, [momentary pause] but none of them seems quite right. [Longer pause] I guess I have some thoughts that just seem so, well, new, or big, or personal. [Longer pause] I have some thoughts. Then I think, well, I'd better not go there. I mean, really, that's too big for the time we have left.

Counselor: So, there are things you'd like to say, but you see that we'd only get in the middle of them and then have to end.

Client: [Long pause.] Well, isn't that true? [Sounding a little frustrated or exasperated] What am I supposed to do? We only have so much time!

Counselor: [Responding in tone to the pain she feels from her client.] I don't know what we have time for, as I don't know what you might say or where it might lead. However, I do hear your frustration with our shortness of time. Also, I do know that the time we have left today and our remaining meeting next week can both be a short time and a long time. You might have time to say and experience much of what you need, but I do also assume we are going to end unfinished.

Client: [Sounding both somewhat frustrated and pained.] So, if it's unfinished, then what's the point? [Taken aback by her own words, beginning to speak more slowly] I'm sorry. I do appreciate your listening to me and caring. I'm just not sure where to go from here.

Counselor: I get that while you want me to know how much you appreciate our time, you are quite distressed over our ending. I gather that you wanted to be at a certain point before we ended. You see that you're not there, and it's scary.

Client: [Very long pause.] Yeah, it is scary. Sometimes, I think my life is a terrible mess. Sometimes, I think I am a terrible mess. Other times, I think it's not so bad. I see myself as strong. I've been through a lot. Okay, I can handle it. I can handle anything.

Counselor: So a real mix, worry if your gonna be okay, and a great knowledge that you will be okay.

Please note that the counselor, in this example, stuck with her client and her client's processing of her own thoughts and feelings toward ending, toward her use of the time, and toward her self-evaluation of readiness. The counselor did not offer false reassurances, though she did offer her limited views on ending. Most importantly, the counselor remained focused in deep empathy and UPR for her client, as she had throughout their time together. The counselor may also choose to talk with this client about where to go from here with her work and development, especially since she seems to be ending in a difficult place, feeling vulnerable.

The Potential for Feeling Raw in Ending Mid-Work Jeff remembers a time when he was making use of a therapeutic relationship in counseling:

> I remember that I had been letting my counselor know of some old hurts that were still there and of my actions that I felt ashamed of. I remember feeling raw from this work, even outside of counseling sessions. An image came to me, when I was walking one day, that it was as if I had opened a wound that would remain open until I had finished more of my work in counseling at that time. It seemed to me that my rawness with this open wound was even evident to others, although I know now and knew then that it really wasn't.

We think that other clients may be experiencing their own rawness in the middle of their work. So, we have worried a little that arbitrary endings might leave clients in such a raw state.

But we have learned that the combination of planfulness around ending (i.e., making your client aware of when the end will be and counting down) and highly attentive therapeutic listening and deep empathy keep this from happening. If your clients fully understand when the end will be, if you are attending carefully to all that they are communicating to you (e.g., what is not said and what is implied), and if you are feeling deeply with them, then your clients will take the opportunity to realize if they are feeling raw or otherwise unready to end, and take the opportunity to discuss that with you and to bring themselves to a readiness to end. They may decide to quickly finish their work before the end or decide to close down and consolidate the work completed so far.

Ending with a "No Show" In settings where clients have full control over showing up for appointments or not, we have had adult or adolescent clients end counseling by just not showing up for the last possible appointment or two. A memory from Jeff's work may help illustrate this.

> I had worked with this young man for a couple of months. Before individual counseling, he had attended a workshop, then a small psycho-educational group regarding career decision making. In our time together, he and I had learned that his indecision was highly personal, as he felt strongly that he knew the career direction he really wanted and believed right for him, but was just as sure that his loved ones would withdraw support—moral, emotional, and financial—if he took such a potentially impractical direction. He was also becoming aware that such indecision was influencing his other life decisions. As our ending had been planful, he was aware of our final two meetings and chose not to show up for those last two meetings. He called and cancelled the second-to-last session, giving some plausible reason, then simply did not show up or call to cancel the last possible meeting. I felt close to him and personally, emotionally invested in his decisions.
>
> After the no-show, I was standing in the break room feeling sad. When I told a coworker, who also knew him, what had happened and told her that I'd wanted a chance to say goodbye, my coworker responded, with caring in her tone, "I guess that is how he said goodbye."
>
> Now, looking back, I wonder if the no-show had been his way of deciding to close his metaphoric wounds over any of the healing that he had completed so far and that he had left to do. Maybe at some level he reasoned that in those last meetings he might not be able to stop from opening himself further, exposing areas that he could not close in time to leave the safety of our therapeutic relationship.

We tell this story in order for you to be aware of and able to prepare yourself for such events. We also want you to know that while you may not know your client's reasons for such an ending, you can know that as long as you have been planful and provided a strong therapeutic relationship, there are reasons, and if you knew them, you could accept them.

The Possibility of Ending with a Big Bang In Chapter 8 we told of Jeff's work with a client who ended by disclosing, as he walked out the door, new information that was personally big and painful to him. We revisit that story in this context:

> From the way it happened, it was quite clear he meant to end our time together with that disclosure. Of course, our ending had been planful and our relationship deeply therapeutic (at least I think so, as this was very early in my work and I have improved since then). So, this was his choice of how to end. Perhaps he decided to lance a wound that had been so far only healed over, to let out some of the infection left behind but to close it immediately back up, before it was too far open to close it back quickly.

While it was shocking to me and I was momentarily confused as to how to respond or to think about this ending, I now see that his way of ending seems to have been a wise choice of how to do the most work in the time that we had. However, if up to me, I never would have guessed or suggested such an ending.

SEEKING FEEDBACK
IN FINAL MEETINGS

Especially early in your work, you may want to seek feedback from clients in final meetings. You will, of course, have received feedback throughout as you have reviewed for client progress, satisfaction, and decisions/intentions toward ending through reflecting what you see. But in the end, your clients may be able to give you feedback with a greater perspective, especially when your work together has come to a natural conclusion. A client-led discussion of your work in counseling, including what it has meant to them, how so, and why, will usually occur naturally when ending with adults and adolescents. If that discussion does not naturally ensue, you can initiate it. We suggest you start by stating your desire for the feedback. For example, "Shelly, I want to know more of how you have experienced your work with me so that I can understand better what it has been like for you and I can continue to improve for my work with others." Then, if your client does not start to offer such feedback, you could foster the discussion with a reflection of what you think it has been like for her: "I have the idea that this work has been hard for you, that there were times when you were aggravated and wanted more from me. But now you are at a point where you see how far you've come. You seem to appreciate your progress, your hard work here, and me." Offering such a reflection gives your client some solid notion from which to let her thoughts and reactions grow, an initial statement to which she may agree, disagree, add to, or subtract from. On the other hand, if you ask it as an open question, "Tell me your thoughts on your experience of work with me and counseling at this time. What has it meant for you? How has it worked or not and why?" you may overwhelm your client with the broadness of the question. Since your client will have done deep, broad, subtle, abstract, and meaningful work in her time with you, such an open question is likely to prompt an answer that underrates this work and reduces it to an answer like, "It was good [shrug]."

Since you know that you are going to want such a discussion in the final meeting and know that it will take up some of your client's time in the final meeting, then let your client know this as you count down. Casually clarify this for your client in your last couple of statements counting down the time near the end, "We have two more meetings after today, and in our last meeting, I will want to take a small amount of time to hear from you what your work has been like with me."

COMMON DIFFICULTIES FOR BEGINNING COUNSELORS AROUND ENDING

Not Wanting to Let Go

Of course you will miss many of your clients. We miss many of ours from years past. Some of our clients have ended their time with us in counseling but continued to live in what we knew to be dangerous or difficult situations. So, we would expect that it is natural for us and for you to worry some over such clients. Fortunately, our worry is not too strong. Our experiences of clients' lives have taught us just how resilient humans are. Especially once the light of self-awareness is turned on, the drive to self-actualize becomes much more difficult to disrupt or divert. This enhanced drive helps persons survive and thrive through dangerous or difficult situations.

But if your anticipation of missing clients or worrying over ending with them seems excessive and to drag you down, especially if it causes you to hold yourself back from relationships, examine carefully (in self-reflection, contemplation, consultation, and in your own counseling) whether or not your warmth toward clients is nonpossessive. If your warmth is possessive, for example if you have thoughts that some of your clients *must* have contact, help, or control from you, then the resulting way of being in you will not only be hurtful to you in ending but to your clients' overall progress and to the usefulness of the counseling you offer.

Our faith in a higher power is important to our allowing our warmth to be nonpossessive. Seeing ourselves as a very small part, albeit a powerful and important part, of a higher power allows us to better accept the limits of our influence.

The Frequent Happy/Sad Endings

For us and for the many counselors we know, most endings with clients are a mix of happiness and sadness. We are happy to see each client go beyond us on their path. We are happy to have the time open again, and then again also sad to see them go. As our connection with each client was real and personal, we feel a hole or void for a while after the client is gone. We miss our clients. That feeling often makes Jeff think of the wistful sadness of fall. He enjoys fall, loves it, but somehow it brings a greater awareness of the passage of time and of missing loved ones who have moved on. It's something like the feeling of clients moving on. It's a good feeling, but sad—a good sadness.

Seeming to Want Too Much to End

In case you sometimes find yourself longing to end with some clients and perhaps feeling guilty or thinking yourself a bad counselor for it, we'd like you to know that this has sometimes been true of us and of many fine counselors that we know. Some clients can be hard to connect with, and so your connecting with those persons can be hard work. The difficulties that some clients experience can be quite painful, and so those emotions are painful to feel with them. Also, sometimes when clients end, it represents a little break from some workplace responsibility.

However, we find that with clients who have been able to continue counseling to a natural ending, we usually have that happy/sad feeling around ending, and while we know it's time to end, we really don't want to end. Those persons who were hard to connect with as clients have usually become much easier to connect with. Those persons whose pain was hard to feel have usually come to experience a fuller range of emotions, even if many of the difficulties of their lives continue.

So if you think you may be overly longing to end with clients too often, please take time to contemplate problems there may be in your way of working. You may wish to consult with a trusted colleague or receive your own counseling for a time. You may need to reconnect with or reconsider some of the ideas, rationales, and reasons presented in Chapter 1 that underlie counseling and review our core skill chapters that follow. We offer a few additional questions for you to consider in such situations. Have you gotten burned out? Are you trying to take undue and unrealistic responsibility for clients? Do you need to find a way to renew your trust in your clients and the self-actualizing forces inherent in life? Do you look to your clients for personal affirmation? Are you trying to do too much?

Client Reluctance to End

In natural endings, clients are not usually reluctant to end. Most often we find that clients feel the same happy/sad feeling around ending, but also a confidence in readiness to end. Sometimes clients may want to continue; sometimes not, as it has been hard work. But even if a client does want to continue, there is a difference between wanting and needing. Some days we might want to stay home with our loved ones all day, but we also enjoy the sense of accomplishment that work brings, as well as the money, so we move on.

If you do find that one of your clients is highly reluctant to end, search your work for errors you may have made in planning to end and providing a therapeutic relationship. Perhaps you have misunderstood this person in some significant way. If you have provided a strong therapeutic relationship, for however short a time, and planned to end well, extreme reluctance to end will be very rare, whether the ending is arbitrary or natural.

If extreme reluctance does occur, feel that reluctance fully with your client and connect with him at that level. If it is possible to extend your time together, let him know that you see his reluctance and suggest adding time for additional meetings. Even if you cannot add additional time for more meetings, it will still be best to fully empathize, accept, and connect with his reluctance. This reluctance must be necessary for this person in his unique situation. So connect and help him to learn from it through your connection, just as you would have at any time throughout his counseling.

Surprise That a Client Seems More
Okay with Ending Than You Do

Sometimes we or other counselors are surprised that a client seems more okay with ending than we are. It may simply be that a client is a less sentimental sort of person than the counselor. It may be that she has connected deeply and used counseling well, but to her when it's time to move on, it is simply time to move on.

It could also be that in ending, this client just does not want to get into her emotions over not wanting to end or missing you. If so, respect this preference, just as you have respected her decisions for what to express and when throughout her time in counseling.

Unknown Reasons for Clients Ending and
the Temptation for Counselors to Speculate
or Blame Themselves for Some Error

This would be more common in agency-like settings, where the adult has more control over keeping appointments or not. We have often found that when clients choose not to continue, beginning counselors find some reason to blame themselves for their client's choice. Some of this introspection could be useful, as there is always room for counselors to improve their connections with clients and the therapeutic relationships they provide. Also, some beginner counselor errors in therapeutic relationships can discourage client engagement. For example, if a counselor is going through the motions of reflecting but not really feeling with and prizing his client, that client might understandably feel judged, misunderstood, patronized, or critical of the counselor and counseling for reasons that she may not articulate. However, we know of numerous cases where the counselor's self-blame was generated solely from not knowing why the client chose to end, often after only a small number of sessions. Following, we offer a few of the many possible happy endings from counseling that often remain unknown to the counselor.

- We know of one client who worked briefly with her counselor, crying hard through her first couple of sessions. She had begun counseling because she was having great difficulty in stopping her crying outside of sessions. After three sessions in which she cried a great deal and in which her counselor felt deeply with her and accepted her tears, she called to cancel two sessions in a row, and it was assumed she had ended counseling. The agency followed up with her, partly because her counselor was worried about her and was riddled with self-blame over the assumed failed ending. In follow-up, she was able to tell the agency that she felt very well, that she had mostly stopped crying, or at least she only cried what she considered a reasonable amount (she did, of course, have some understandably painful difficulties). With this information, the counselor was able to theorize that her accepting her client's pain and tears, while

feeling the pain fully with her, helped this client also accept her pain and
tears. Then, as this client stopped fearing and fighting her pain and tears
so much, she found that they lessened and weren't so strong. So, there was
a happy ending for the client, but until knowing this, the counselor had
blamed herself and assumed an unhappy ending. Remember this in the
many situations in which you will not know for sure the outcome of your
work with clients you serve.

- We know of numerous clients who came for only one meeting with their
counselor and found that one session with the opportunity to freely self-
express, to be heard, felt with, and understood, was enough to sustain
them and recharge their movement along their path of self-actualization.

- We know of many clients who have made a solid start in counseling in a
small number of sessions, then decided to end for a while, and returned to
work in counseling later when they felt ready to move into their next
phase of work. We know of some clients who have made such decisions
quite openly, actually telling their counselors something like, "There really
is more that I have to say [implying that it's big], but I just can't go there
yet." However, opportunities for such openness are rare. Because of Jeff's
work overseeing counseling interns in a setting over a period of years; he
has had the opportunity to see clients who had ended with their original
counselor return to pick up their work later with their next counselor.

There are countless other endings in which there is no clear knowledge of
why the client chose to end when she did. In too many cases the counselor's
tendency was to assume the worst and blame herself for failing. An excellent
remedy for not knowing is well-planned, systematic, but simple research into
client satisfaction and counselor effectiveness (such as client satisfaction surveys,
pre- and post-measures, and comparisons with similar groups who do not use
counseling). Such research cannot determine a counselor's effectiveness with
every single client, but it can provide evidence of a counselor's general effectiveness and give insight into how the counselor might improve.

GREAT SATISFACTION
AND JOY IN ENDINGS

Remember that great satisfaction and joy in endings can be the norm, rather
than the exception. As you build and maintain your therapeutic relationships
well with clients, you can experience great satisfaction and joy in your clients'
progress and development with them. As you build and maintain therapeutic
relationships with your clients, you help your clients build a process of ongoing, self-aware self-actualization and self-aware decision making. This is a
process that easily snowballs to become a way of living for each person. It can
be growing strongly after meetings with you have ended.

ACTIVITIES AND RESOURCES
FOR FURTHER STUDY

- Journal, discuss, and essay your explanation of how the principle of independence is woven throughout the skills of therapeutic relationships.

- Working with a peer study group, create a variety of counseling scenarios (different types of people, different kinds of concerns for those persons, different life situations, natural and various arbitrary endings). Plan together how you might structure a planful ending for each different situation, why you would structure it that way, and how you think it might play out, including both the most-likely scenarios and possible problematic situations bringing up difficulties in ending. Role-play aspects of these scenarios, such as final sessions. Practice some problematic situations, but don't over emphasize them. Remember that the strength of the therapeutic relationships you provide will prevent most problems in endings.

- In consideration of our section, "Recognizing the Many Forms of Progress," discuss with peers and/or contemplate alone the many more forms that progress can take that we have described so far. Think in terms of internal, personal changes in clients and then new behaviors that would likely be natural and durable outgrowths of those internal changes. Consider clients that might use a variety of different counseling settings and clients that bring a variety of initial concerns in order to generate an understanding of as many different forms of progress as reasonably possible.

- As it relates to being able to tell a client how you see her in a final meeting and throughout her therapeutic relationship with you, explore the issue of feedback with a peer group. How is giving feedback that you expect will be positively received different for you and others than feedback that you expect will be negatively received? Contemplate, journal, discuss, and essay your thoughts on that question as well as how to tell clients the ways you see them that may be received positively and those that may be received negatively.

- In a safe atmosphere, such as in group counseling, practice giving your impressions of others that you think may be important to them, both those impressions that may be received positively and those that may be received negatively. Also, seek such feedback for yourself. Contemplate, journal, and discuss your immediate, then longer-term reactions to giving and receiving such feedback.

- With a group of your peers, generate counseling scenarios, some of final sessions, some of earlier sessions, where you role-play telling clients how you see them. Include perceptions that you expect will be received positively and those you expect to be received negatively. Remember that when you are telling clients how you see them in final sessions, what you say should not be news but review.

- Carefully consider difficulties and strengths you may have in letting go of clients whose struggles you know will continue after their time with you in counseling. Carefully consider difficulties and strengths you may have in letting your warmth for clients be nonpossessive. What strengthens you in these endeavors? What makes your work more difficult in these endeavors? Can you find beliefs (spiritual or otherwise—i.e., about yourself and your relationship to your world) that strengthen or weaken you in these endeavors? Contemplate, journal, discuss, and essay your findings on these questions.

- Consider the case of a counselor who longs too much to end with clients. Whether you imagine that this person might be you or a colleague that you care for, review the concepts of this book in seeking to discern the possible problems in a counselor's ways of working that produce this overlonging. Discern the possible difficulties these problems might be causing in this counselor's therapeutic relationships. Then consider ways that this counselor may be able to work through some of the most-likely problem areas. As with so many of our activity suggestions, you can contemplate alone—take the time to think, journal, and essay in a disciplined way. However, it will be even more productive to further develop your ideas and learn from those of others through discussion and ongoing contemplation.

- Following the same guidelines suggested for the preceding activity, explore the potential meanings for a counselor's work if he frequently finds that his clients are often highly reluctant to end. Also explore the possibilities of events in clients' lives that might be prompting their reluctance to end.

- Search for and review various examples of research that seem simple and provide evidence of counselor effectiveness. Share the examples you find with your peers. Plan how you may employ similar research in your work settings in order to help settle your mind over the effectiveness of your work and help you discern areas to improve. Periodically implement such research in your practice in order to confirm your success, build your confidence, and shed light on areas in which you can improve.

- For the first clients to whom you provide significant therapeutic relationships, essay then discuss with peers (within the limits of confidentiality) how those clients have changed and what aspects of your relationship with them seem to have affected that change. Make clear the qualities or behaviors that have changed for each client or how his or her life has improved and/or is improving. Explain what it was about your therapeutic relationship that helped your clients change or that benefited them. Be as specific as possible in describing the mechanisms or aspects of your therapeutic relationships that brought change or benefit to each client. In individual cases, you cannot prove what it was that helped clients change, but your task here is to articulate what you think it was about your therapeutic relationship that helped your clients change and to clearly present evidence that leads you to your beliefs of what helped your clients change. Repeat this exercise periodically throughout your career.

- Revisit the focus activity for this chapter and consider how your thoughts and answers have changed or been reconfirmed.
- Review the Primary Skill Objectives for this chapter. If you have not yet mastered them to your satisfaction at this time, please reread, engage in additional practice, and seek additional readings and study opportunities until you have mastered them to your satisfaction.

12

Therapeutic Relationships Across Cultures

If we are to achieve a richer culture, rich in contrasting values,
we must recognize the whole gamut of human potentialities
and so weave a less arbitrary social fabric, one in which
each diverse human gift will find a fitting place.

MARGARET MEAD

PRIMARY SKILL OBJECTIVES

- Be able to explain why counselors must reach out in order to serve persons that are or seem to be culturally different from themselves.
- Be able to offer several ways that counselors can reach out and make the counseling services they wish to provide seem useful and valuable to persons who may have had understandable reluctance to seek counseling from a person or persons who appear to be culturally different.
- Be able to explain the importance of each core condition of counseling in reaching across cultural differences once counseling has begun.
- Be able to explain how meeting clients with humility can be beneficial in counseling, especially when the counselor and client are culturally different from each other.

- Be able to explain several of the ways that immersing yourself in cultures that seem different to you can benefit your understanding of those cultures, other cultures in general, and yourself and the culture that you identify with or are most comfortable in.

- Be able to explain what we mean by "thinking broadly" regarding cultural differences and add your own examples to ours.

FOCUS ACTIVITY

The following exercise requires you to imagine yourself in the role of a client in a particular situation. Please imagine the scenario as fully as possible in order to understand what this client's experience may be.

Imagine that you have decided to use the counseling services available at your college, school, or community. This imagined scenario may take place at the age you are now or at a time in the past. Try to create a clear mental picture of why you are seeking help. Whatever your reasons for counseling, know that your reasons likely involve some emotional distress or pain, as without distress or pain, most persons will not seek counseling. Imagine that you made the appointment to begin counseling without actually meeting the counselor you will work with. You are given an appointment time, but that is all the information you have at that time. Now imagine that between the time of making the appointment and beginning counseling, you start to form a mental picture of what counseling may be like and what your counselor will be like. In your mental picture, do you see that your counselor has had similar experiences, will easily understand you, has had a life much like yours, and is on a parallel course, just a few steps or stages ahead of you? Or have you begun to think that nobody's experiences are like yours and that there is no way that this counselor will be able to understand you? Alternatively, you may have a hard time picturing the counselor at all. As you create this scenario, create the mental picture that is most like what *you imagine* your expectations would have been or would be.

Now put yourself in the position of arriving for your first appointment with your counselor. When you arrive, you see that apparently your counselor is quite unlike you. The counselor is not dressed as you expected or as you would dress. The counselor is not the age you expected, appearing to be much older or younger than you. The counselor is of a different gender and seems to be of different backgrounds in ethnicity and/or sexual orientation from you. Imagine meeting a counselor that upon first introduction is apparently very unlike you.

Now take a moment to picture and put yourself as fully in this scene as possible. Fill in details in your imagination of what the place looks like; what the counselor looks like; how you feel physically; how you feel emotionally; what you are thinking about counseling, the counselor, and yourself; what feeling seems to be "in the air" (i.e., tension, anxiety, calm); and how you are and/or imagine you would be responding to all of this.

Now write or discuss with partner(s) your answers to the following questions, then keep your answers in mind throughout this chapter:

- What is your visceral reaction to the situation, that is, what do you feel physically (e.g., suddenly at ease, tense stomach, sweaty palms, energized, fidgety, hot, achy)?
- What emotions do you feel?
- What thoughts do you have of or toward your counselor?
- What do you now imagine counseling will be like, and to what degree do you expect it to succeed or fail?
- What may help you come to be most ready to work with your counselor? What qualities in this counselor would you need to feel comfortable in relating?
- What would not help? What qualities in this counselor would make you feel more uncomfortable in relating?

If you have not already experienced this personally, imagine how difficult it might be to seek counseling with a person who is or seems culturally different. You might naturally expect, at least beneath full awareness, to be prejudged and misunderstood. You might expect just communicating what you mean to say to be slower or more awkward, and it may be.

INTRODUCTION: ALL COUNSELING IS CROSS-CULTURAL—BUT YOU HAVE TO REACH OUT

We lived overseas—on the island of Guam—for a couple of years, where we were among a small minority of foreigners. Even though we intentionally chose to live and work in a more rural area of the island among persons local to the area, we would periodically realize that we had unknowingly separated ourselves and that we spent much of our time with other persons from the United States—persons who, like us, were from various states on the U.S. mainland and who had come to Guam as teachers, counselors, or other professionals but who had not grown up on or were not local to the island. Each time we realized what we had done, we'd increase our efforts to more fully integrate with the local cultures and traditions.

This experience continues to serve as a lesson and reminder for us of just how hard it can be to comfortably connect with persons who are culturally different. For us, at the time, it was just easier and more relaxing to be with other people from the United States—we understood the same jokes, used similar expressions, and we didn't have to *deal with the discomfort* of being culturally different or "on display" as foreigners when we were together. Though we attended church, worked in schools, and took part in many wonderful traditional activities with cultural groups local to the island, we were always aware that we were not part of the main culture and background on Guam. We were not local to the island, and we were in the minority. Though we were often made to feel included and welcome, we were also at times reminded that we were "outsiders" and unwelcome. It was a good, sometimes painful, and often humbling experience for both of us.

Interestingly, it was during our work hours there—as a school counselor and a school psychologist—that we were best able to reach out, relate, and build relationships with people local to Guam. During our work hours, we spoke with, joked with, cared for, ate with, listened to, played with, laughed with, cried with, fought with, and fought for a diverse group of clients and coworkers of various ethnicities, ages, lifestyles, and backgrounds. We interacted with grandparents and parents, children and teenagers, teachers, school staff, and administrators. During these work hours, we rarely felt in the minority, and we often felt the common bond of the human struggle and the healing power of human relatedness.

Indeed, it became apparent to us that the *nature* of our work as counselors was what allowed us to become more "the same" and feel more at home with the people of Guam. In the striving for our own best work as a school counselor and a school psychologist, we were both able to reach out—to share ourselves as unique and caring human beings who were willing to connect in a genuine manner with respect for other unique human beings.

We found this connectedness of shared experience and human relatedness to be most apparent and powerful in the language of play and art shared by children in counseling sessions. It is in our counseling of children and providing child-centered play therapy that we have most often been reminded of the importance of the providing the core conditions and allowing each client to set the pace and use the language most natural for self-exploration and self-expression. When a child is given the opportunity to experience his true essence and communicate freely without fear of reproach, punishment, or judgment, it becomes possible to glimpse the real and wonderful potential of diversity. We have found no better way to affirm our faith that *each one of us is unique*—each human being an unrepeatable miracle—than to facilitate such self-exploration and self-expression in therapeutic relationships.

To us, all counseling is cross cultural if focused in the core conditions and skills of therapeutic relationships. The whole point of counseling with the core conditions and deep therapeutic relationships is to experience with another person, to enter the other person's world and allow that person to learn from her experience with you and your expression of your experience with her. Revisiting the meaning of just one core condition, empathy, remember that it has been described as walking a mile in another's shoes. Also, Rogers (1980) explained, "[Empathy] means temporarily living in the other's life, moving about in it delicately without making judgments" and "[t]o be with another in this way [empathically] means that for the time being, you lay aside your own views and values in order to enter another's world without prejudice" (p. 142–143).

However, just as you might provide a comfortable, safe space for a dear friend in need, you will want to explore your ability to create a safe, comfortable space for a diversity of clients. This is indeed part of showing respect, empathy, and warmth. For example, you have to make your practice accessible to persons who may be culturally different from you in order to have the opportunity to reach across cultural differences with the core conditions and your skills in deep therapeutic relationships. You have to become a person in a place that those who sense themselves as culturally different from you can feel reasonably comfortable and accepted.

There are many ways to reach out and be more approachable. Some counselors have surveyed communities regarding services that those communities might want. This has the effect of suggesting services that are possible and letting the community know that you are there and interested. Presentations on the services you offer and mental health-related topics are always a standard and effective way to reach out. Often it is helpful for the people you wish to reach to see you humble yourself and even be a little nervous in presenting to them, to see that you are not so unusual or intimidating—but just a caring person who really wants to be of service. At times, if you can make one professional friend with knowledge of a community you wish to serve, that person may be able to point you to key individuals in that community who may help you broaden your contacts.

REACHING ACROSS CULTURAL DIFFERENCES WITH YOUR SKILLS AND YOUR SELF

Once your client has come in for a first appointment, your real reaching out is with the core conditions that you have developed in yourself and with all your skills in forming therapeutic relationships. It is impossible to match techniques to persons or populations, so best to focus on *ways of being* in counseling that reach across cultural differences. To us, that way is through using the skills of therapeutic relationships. Empathy means you are trying to come as close to understanding that individual as possible, which includes his unique set of cultural experiences. UPR means/necessitates that you accept and prize that individual and the person that his cultural experiences have helped shape and develop. UPR leaves *no room* to prejudge cultural groups or value sets.

Genuineness means both that your empathy and acceptance of the individual are real, and also that you know what you know. You bring yourself into the counseling relationship—meaning that you are not an amoeba of various counseling techniques, ready to change and/or use a technique depending on the culture or age or socioeconomic background or sex of your client. Genuineness with, empathy for, and acceptance of each individual client allows you to use what you know about counseling and connecting with people to reach a diverse group of clients.

HUMILITY

Instead of merely studying or reading about issues of diversity, make it your job and way of life to learn about and experience persons who may have differing sets of experiences from you and your cultural background. Realize what you don't know and that it will be impossible to ever know it all, but strive for that goal just the same. Perhaps more importantly, strive to meet each client with the humility of knowing that you don't know what her life and cultural experience is like for her, even if you think you know what cultural group that person belongs to.

Be Wary of Cultural Assumptions

Jeff had an experience in counseling that has helped him remember not to assume he knows another person's cultural experience.

In reference to a current situation and other life struggles, a client happened to say, with mild disdain, something like, "You know what it's like to grow up Baptist in these mountains." I think he actually meant, "Everybody knows what it's like to grow up Baptist in these mountains," using the stereotype for a shorthand explanation of his experience. But his statement struck me because from my background, I probably do know a lot about what it's like to grow up Baptist in the mountains that he was referring to; at least I know a lot from family experience, and I still feel emotions related to some of my early experiences there. So, my first impulse was to respond something like, "Yeah, I sure do," or "I guess I do [implying understatement]." However, this client was just beginning to let me know something about him and his life, so my second impulse was to stop my first. I realized that I really did not know what it was like for him and may have been making dangerous assumptions from my experience. I realized that my understanding of what it was like for him was important to his story and that there seemed to be something that was important to him for me to understand about it. The reflection I stumbled into came out something like, "I might. I'm not sure. There seems something of your experience that's important for me to understand." So then he took the time to tell me what it was that he wanted me (and perhaps himself) to understand about that part of his experience. As it turned out, I think his experience was very different from mine. That moment in my work helped me to remember not to assume I understand another person's cultural background. One might have assumed that while his and my cultural backgrounds seemed quite similar, I could assume things about him. We appeared to be of the same ethnicity, gender, regional and religious background, and even somewhat the same age. But that didn't make our unique, individual experiences the same.

Sometimes we think it is easier to meet a client with humility when the client and counselor appear to be very different. We find it important to meet each client with that same humility whether that person appears quite similar or quite different.

Know Yourself through Immersing Yourself

A part of being genuine is knowing the values and prejudices that you bring to each relationship. No one is completely open, a blank slate. If you were, you would not be an effective counselor, as you would not be a real person.

So, you must work to know yourself. Contemplate what you have learned from your own culture, from your family, your communities, your religion or spirituality. Deeply contemplate how you have learned this. Read and study other cultures, but be careful not to take the scientist/observer approach, like looking at bugs under a microscope. Travel and immerse yourself in other cultures. Don't visit them like Disneyland, where you only view representations of other cultures but truly immerse yourself. The purpose in such a major life endeavor is not only to help you understand the other cultures you immerse yourself in but also to help you understand your own by experiencing cultures that are significantly different. By removing yourself from the culture and lifestyle you identify most closely with, you are then better able to understand those clients who seem most culturally different from yourself. By immersing yourself in another culture for a time, you will be able to view your own cultural expectations and attitudes in a way you may not have previously. Stepping away and getting a new perspective is very likely the best way to have the clearest view possible of your own cultural background. Finding yourself in unfamiliar territory and taking notice of your own fears and defenses is very likely the best way to develop empathy for all your clients.

THINK BROADLY

Don't narrow your expectations for persons whose cultural differences may be significantly different from your own to include only persons who seem to be of a different gender, ethnicity, and geography. The concept of human diversity is both complicated by and oversimplified by the many attempts to define and understand it. Jeff finds himself quite different from some of his relatives, even though they share the same heritage and can be considered to represent some of the same cultural categories. Indeed, many of us have grown up to have very different values, interests, behavioral norms, and worldviews than our relatives or main culture of origin. We remember a coworker, who is African American, who seemed to feel great irritation when it was assumed that she would naturally understand the experience of students who were African American and from a nearby large city, when she was from a small town in a rural area. The irritation she felt was in large part due to the assumption based solely on her skin color that she would be the best counselor for African American clients.

Nancy remembers working with two young boys—both had lost their fathers in the war in Chechnya and were visiting the United States in a program for orphans of war. Because the boys had this common experience and were in a foreign country and community, they were indeed connected by this experience. But, at the same time, they were two unique and very different little boys. During a counseling session, one of the boys grew frustrated and angry that others assumed he was in any way "like" the other boy, and he exclaimed, "He [the other boy] is nothing the same . . . he lives in another place . . . far . . . far . . . many miles . . . from me in Chechnya . . . he is not like me and my family!" It was apparent that this boy was asserting and holding onto his individuality, and in need of doing so in a foreign country and community. Though he had some things in common with the other boy, he did not want to be confused with him or to lose his sense of individuality.

By thinking broadly, as counselors and helpers we can hopefully represent and relate to those individuals who may feel misunderstood and restricted by stereotypical viewpoints, assumptions, and prejudice. When we were foster parents, we remember persons who felt no hesitation in making remarks or giving us advice about our African American foster daughter. Because we are Caucasian, we always stood out whenever we went out as a family with our foster daughter. As is many times the case when talking about children, remarks were often made loudly and in front of our foster daughter with little respect for her feelings. One of the most memorable remarks, and perhaps the least respectful toward our foster daughter, was made on several different occasions by persons of both races. It always seemed that those who said it meant the remark as a compliment of sorts. "There will be a special place in heaven for you," they would say in reference to the "reward due" us for having taken our foster daughter into our home. Though we are both Caucasian, we both grew up in rural southern areas and attended rural southern public schools. Jeff and our foster daughter both grew up attending southern Baptist churches. We shared with our foster daughter an understanding of the rural, southern, sometimes very painful, and lasting attitudes about social class, race, and poverty in the South. All three of us understood that in some ways this thoughtless remark was simply "southern." All three of us also understood, each time it was said, that it carried a hidden message that was hurtful to us all but mainly to our foster daughter—an implication that she was "trouble" to us and that we were due a reward for the sacrifice we were making.

Still, we also shared something more. We always seemed to feel great empathy for one another during these uncomfortable moments. Most of the time we were all just trying to enjoy the moment eating our pizza or taking a walk on the beach together. When people would intrude with remarks, it seemed that the three of us began to share an awareness that some people sometimes say too much. Knowing this, and somehow coming to expect it, seemed to help us feel more "familiar" and bonded to one another. Thankfully, we were also able to talk with our foster daughter openly, and listen to and accept her feelings about instances when she felt uncomfortable, sad, angry, or ashamed about being with us. We were able to laugh and talk, pray and eat, sing and cry, dance and play together. We seemed to have a lot in common as human beings. Her 9-year-old thoughts and remarks on the issues of race, poverty, family, and life in general were always refreshingly honest, wise, and to the point. She certainly helped us to think broadly!

The Fairly Foreign World of Children

Most adults don't engage in pure fantasy play. If we were seen doing this, we might be thought insane. Role plays are acceptable, daydreams are acceptable, spinning stories with words is acceptable, creating with art is acceptable, acting in a play is acceptable, but just pure fantasy play certainly is not. The vast majority of our adult communication is with words, but as Axline (1947) explained "Play is the child's natural medium of self expression" (p. 9). Even when adults go to a gym, which in childhood may have been an indoor place to play, we "work out;" count pounds, miles, minutes, or calories; or tend to play goal-oriented competitive games.

Children, when freed from requirements to meet expectations or perform for adults, play naturally and from the pure core of their being. They use this play for emotive release during stress, for resolution of problems, and for mastery of life situations. Nancy has often explained to students that it is in working with children that she finds it possible to see the real potential and beauty of human diversity. It is in working with children that both of us have built our faith in human beings and have truly seen that each individual child has her own unique voice, language, and style of communicating. It has been possible to extend this faith in human beings and in the unique individual when working with older clients. We remind ourselves that "words get in the way" sometimes, and that our older clients are often restricted, feeling guilty and afraid to fully self-express. We remind ourselves of the power of the therapeutic relationship— of being a safe and freeing place where both children and our older clients will experience the warmth and acceptance necessary to be themselves.

A child does not have the words, or at least does not have the cognitive-emotional integration *to use* words to work on such existential concepts as seeking self-identity and acceptance of it, discerning what he believes his relational potential with others is, and what he would want it to be. But children need to and do work on these concepts. In fact, such concepts are set by the end of childhood and the beginning of abstract thought. Such concepts are set even *before* a child has the *ability* to examine core beliefs in the same ways that adults would. Core beliefs are, in effect, established, confirmed, and set in darkness. If those core beliefs are errant and troubling, they are much harder to change than they would have been before becoming solidified. So, when events or a child's perception of events have damaged or are blocking his path to self-actualization, that child can benefit greatly from counseling focused in a therapeutic relationship. This requires, however, that the adult find a way to reach across the formidable divide from adult thought and expression to the child's innocent and pure expressions in actions and play. This is indeed a "foreign world," but as discussed previously, it is a world wherein the skills and the core conditions for building therapeutic relationships are the necessary bridge between these two worlds—the worlds of adult and child.

Working with children requires an understanding of and respect for children as a diverse population—diverse as unique individuals but also as a whole population that is different from the adult population.

Clients and Others Who May
Not See the Value of Counseling

Among counselors and mental health professionals, we form a culture that holds similar sets of beliefs about the importance of self-awareness, self-responsibility, communication, human connections, and other counseling-related concepts. These concepts can come to seem self-evident to those within our culture. Yet, such concepts may not be self-evident to larger cultures. Jeff remembers a momentary debate about "just what Jeff does for a living" among his elders at a family reunion:

> Some of my aunts and uncles were discussing just what I did for a living, and there seemed to be some confusion. One of my great uncles teased me, with a wink, and said something like, "You ought to be one of those doctors where people just come to tell you their problems and you just tell them what to do." One of my cousins, who has a pretty good understanding of counseling and what I do for a living, quietly answered, "I think he is."
> I realized from her look of sympathy in that moment, and from my own common sense, that I had no real chance to explain what I do for a living to my elders in that moment, even though they love me and might be interested in understanding. I had no real chance to explain that counseling is *so much more* than just telling people what to do. But the explanation would take too long and also put a damper on the fun of the day. So, I let my great uncle's teasing explanation of what I do for a living stand. I realized the gulf of differences in our experiences and background knowledge. While we love and respect each other, our background of experiences, in this case for understanding counseling, have been hugely different. That moment also stands as a reminder to me to continually redouble my and my students' efforts and abilities to explain to clients, loved ones, teachers, administrators, or third-party payers just what counseling is and how it helps our clients and our society.

Another cultural gulf for counselors and mental health providers from related fields is presented by the predominant medical model mindset of our society. In our travels we have realized that Americans (including ourselves) have a well-deserved reputation for impatience. We tend to hope for "quick fixes" and "magic pills." Truly, our medical doctors are able to do amazing things, helping us all live longer and better. Because medical care has provided us so much, our society may have come to rely too heavily on and to have unreasonable expectations for its power. For example, the onslaught of advertisements for medical drugs inundates us with a sometimes subtle and sometimes not-so-subtle implication that if one finds the right pill, all will be well. We have realized from the implications of these ads that apparently there are pills to fix all our moods and troubling thoughts. The ads that we have seen don't seem to mention that a person experiencing depression might also need to make life and self-changes to feel well. Apparently there are pills to improve men's sex life or penis size that would then cause them to be more confident and competent and to be seen as different—suddenly changed for the better by all those who they meet. Apparently, there are pills, easy diets, or medical treatments that can make all women more beautiful and cause them to automatically feel better about themselves and their lives.

As a result, explaining counseling and beliefs common to counselors can be difficult. Self-development, change that may take time, great effort, and be emotionally painful, is a tough sell in a society that advertises "quick fixes." Most any thinking human would doubt the assertions of the advertisements that we mentioned having seen when the assertions are stated outright, as we have done. But the assertions aren't stated outright; rather they are an ever-present, emotionally manipulative undercurrent in our society. They assert commercial interests by relentlessly feeding voice and imagery into an undercurrent that appeals to our hope for simple solutions and a tendency toward linear, straight-line thinking. It can be easy to assume that the quickest route from point A to point B is a straight line (anyone who has lived in mountainous areas may tell you this is often not true). When linear thinking is related to the work of counselors, one could easily think that if behavior is the problem, the solution must be behavioral. We find that this is at most a partly true assumption. If behavior is the problem, a behavioral solution may help but would be greatly enhanced if the emotions and thoughts that drive the behavior or the complexities of the behavior itself were more fully understood, explored, and expressed by the persons for whom the behavior in question is a part of their experience.

It is important for members of the culture of counselors to remember that while counseling solutions or applications seem natural to counselors, members of the culture of teachers more naturally gravitate to teaching solutions or applications for the same problem. Likewise, administrators more naturally gravitate to administrative solutions or applications to the same problem. Quite possibly, some combination of the approaches may be most helpful. But now that we are considering combinations of approaches, life is getting complex again. And working to implement complex combination solutions in a society that wants quick, easy fixes will tax both the explaining and the patience, listening, empathy, and UPR of the counselor who wants to make it work.

Because concepts that become second nature to counselors and mental health professionals are usually not second nature to persons outside of our culture, each counselor and mental health professional must be ever vigilant for how the assumptions of our culture about human nature, problem solving, counseling, and related concepts affect our thoughts, actions, and communications. Those from our culture must be ever vigilant to translate our culture's beliefs and actions into terms acceptable and understandable to the larger cultures. For example, when you advocate for funding and support for an expansion of counseling services, put your request in terms of the goals of the community, with a practical, understandable explanation of how counseling will help achieve those goals.

The same goes for explaining counseling to a client. Generic explanations and platitudes may be only minimally helpful. But explaining the potential of a client's use of counseling, related directly to her life and in terms that relate to what she has told you about herself and situation will help.

COMMON PROBLEMS OR EXPERIENCES OF BEGINNING COUNSELORS IN COUNSELING ACROSS CULTURES

Opportunity to Experience a Diversity of Clients

A majority of beginning counselors we know have been middle-class white women, and a large number of their clients seem to also be middle-class white women. Yet, persons from wide varieties of backgrounds can and need to benefit from therapeutic relationships in counseling. Prepare yourself to reach out to needful and underserved persons in our societies by seeking many opportunities to counsel persons with obvious cultural differences from your own early in your work, especially while you still have maximum supervision to help you in this work.

Difference as Advantage to the Counselor

Some of the beginning counselors we have known have explained that counseling persons who seemed quite different from themselves early in their work was an advantage in that it kept them from the habit of making assumptions about clients' experiences. Conversely, we have known beginning counselors to struggle not to assume that their client's experience is quite similar to their own experience when serving clients who seemed very much like themselves. So, again, seek experience with clients who seem to be different from you and remember not to assume that your experience is your client's experience, because considering the diversity of our individual human experiences, odds are your two sets of experiences actually have a great many differences.

Missing the Feelings for the Cultural Context

Some beginning counselors have explained that in trying too hard to understand the cultural context and differences of clients who appeared quite culturally different to them, they missed connecting through feeling with those persons. We see that it works the other way. If counselors first connect with each client's feelings, which are core to our humanness, then it becomes easier to cognitively understand that person's cultural context and differences.

Missing the Content for the Context

Likewise, some beginning counselors working with clients who appeared quite culturally different have missed getting the gist of the content of what their client was telling them by getting caught up in the intellectual task of understanding the client's very different-seeming world. This is especially true when neither were native speakers of the same language or they spoke the same language with very different dialects. There again, it is most helpful to connect on a feeling level. You can know a person's feelings even without words. Then, if you connect with the person through feelings, more of the words will become obvious, and understanding every word will become unnecessary.

Significant Value Differences

Beginning counselors that we know, and we ourselves, have experienced the greatest difficulties connecting with clients who have significant value differences. For example, one counselor we know, who was on her own from an early age and is a largely self-made success story, had great difficulty understanding a client who seemed overly attached or enmeshed with her family and struggled to separate or individuate. We and other counselors we have known have struggled to understand clients who have abused children, spouses, or other weaker persons in their lives.

We find it important to realize that such difficulties in understanding and connecting are difficulties in accepting and empathizing. We remind ourselves that it is that very acceptance and empathy that may help the person who has values that are very different from her counselor's move on to what we assume to be more mature and optimally self-actualized values. For example, if a woman who seems enmeshed with her family needs to individuate, experiencing her counselor's acceptance and empathy will help her more fully feel what she feels and, in accepting her feelings and herself, come to see her situation most clearly and move to change her course. A client who seems less than genuinely repentant for abusive actions may be working hard to keep the meaning and ramifications of his actions out of his full awareness. If so, his experience of his counselor's empathy and acceptance (which does not mean agreement) will help him fully admit his actions, thoughts, and feelings to himself and thus move to change. However, if his counselor overtly or covertly chastises him for his apparent lack of remorse, then he is likely to be prompted only to dig himself deeper into a hole of denial.

The Experience of Connecting

We would like you to know that overall our students do not experience great difficulties connecting with clients who are culturally different. Many times over, we and our students have found great joy and excitement in connecting with many different clients. We are confident this will also be your experience as you strive to provide ever-closer and deeper therapeutic relationships for your clients.

Sensing a Need for Information and Context Education

As we have worked with groups of beginning counselors, we have often had the experience of one member of the group letting the rest of the group know that she feels uncomfortable with her lack of knowledge of the cultural context of a client who seems very different from herself. We admire the counselor's self-awareness and humility in realizing this and expressing it in a safe group of peers. From this disclosure, often a group discussion follows that first focuses on how this counselor feels with the client, then gravitates to a discussion of how this counselor sees the client as different. This, of course, requires the counselor to discuss her own cultural background, values, and biases, as they relate to the client. So, it is not just a discussion of the client but yet another experience for the counselor to learn more of herself. Usually after those initial areas of discussion comes a sharing of perspectives, information, and understanding of the two cultures in question (the counselor's and the client's), including relevant professional literature that members of the peer group may suggest.

Reaching Out and Becoming Accessible

A problem that we see for some beginning and experienced counselors is reaching out and becoming accessible to persons who are or sense themselves as culturally different. It is the task of every counselor to reach out to the communities she serves. If the community doesn't know you are there, see you as approachable and useful, and think that the counseling you offer might be helpful, then why would persons in the community seek counseling? Many persons who gravitate to the counseling profession can be introverted and might prefer to shy away from reaching out to the communities they serve. If you combine that understandable shyness with looking or feeling different from many persons in those communities, then there is a danger that you might not reach out.

It may help to remember not to worry if you look nervous and fumbling in your attempts to reach out. This may serve to make you come across as real and gutsy for being nervous and trying anyway. Jeff is often shy in these tasks. Public presentation has not always been his strength, and making small talk at social gatherings certainly is not. But he has found that when he finds ways to join with others in community projects, he doesn't think so much about presenting himself, and others usually come to like and respect him. We'd also like to offer a story of Jeff's experience with reaching out as a new elementary school counselor, while also suddenly immersed in a seemingly different culture:

> I got my job as an elementary school counselor by accident. I had never meant to work with children. They seemed foreign to me. But when we were moving overseas, the contract I was mistakenly sent was at an elementary school. I told Nancy, "I don't know much about children." I had always been an "ideas and behavior" guy, great with troubled youth but definitely made for adolescents or adults. But we were too poor to afford the overseas phone calls to have my contract changed. So while quickly learning to adapt my skills for children, I also had to learn how to reach out to them. They didn't get my jokes or understand some of my words.
>
> I found that the most effective way for me to reach out to kids was through play and following their lead on the playground and in other free play moments. I decided not to try to talk or present myself (although I did classroom presentations that were reasonably competent but nowhere near great). Whenever I could make a free moment, I immersed myself in play with the children of my school. I focused on one child or one small group at a time and joined them in play in their ways and at their level.

I probably looked silly to some adults (no one else joined in free play this way), but soon I became like "Elvis" to the under-10-year-olds in my village. Walking on the playground, I would often be surrounded by throngs of adoring "fans." It was as if word spread among the children, "He gives undivided attention! [Rather than just instruction, like so many adults.] He likes us as we are! [And didn't come to judge us or teach us to be better, like so many adults.] He thinks we're cool." Now don't get me wrong, I remained an adult. I knew when it was time to play and when it wasn't. But when it was time to play, I played.

My play therapy became highly sought and prized. My free play, combined with my presentations to children and school adults about what counseling was really about, quickly began a flow of eager referrals. As the community of adults saw the effectiveness of my work, they came to respect and value me more and more.

Our point is that there is always a way to reach out and begin a connection, even for shy people. So, we encourage you to let yourself go and free yourself to find a way that works both for you and for the clients and communities you wish to serve.

ACTIVITIES AND RESOURCES
FOR FURTHER STUDY

- We invite you to challenge yourself to a thorough and tough essay-writing project that can help you in knowing yourself in important ways that prepare you for counseling across cultural differences. This essay can be a major part of discovering what your values and beliefs about human nature are, and the origins of your values and beliefs. To give your learning through this essay the energy and attention it deserves, it should be difficult and time consuming. You should work through many drafts and revisions, considering and reconsidering both your ideas and your writing. Guidance and specific instructions for this essay are found in Skill Support Resource D.

- Revisit the Focus Activity for this chapter and consider how your thoughts, feelings, and answers related to the situation(s) you imagined have changed.

- We recommend you read and consider Glauser, A. S., & Bozarth, J. D. (2001). Person-centered counseling: The culture within. *Journal of Counseling and Development, 79,* 142–147. We find it offers a helpful perspective on counseling across cultures.

- We also recommend you carefully review the rich and perhaps provocative views expressed in a pair of articles by C. H. Patterson. In "Multicultural Counseling: From Diversity to Universality" (*Journal of Counseling and Development, 74,* 227–32), Patterson provides an excellent review of the literature on multicultural counseling and problems therein, then asserts four universal elements that, if developed in counselors, create a bridge to more effective counseling across cultures. In "Do We Need Multicultural Counseling Competencies?" (*Journal of Mental Health Counseling, 26,* 67–73), Patterson describes faulty assumptions underlying attempts to specify skills for counselors serving clients of different cultural backgrounds and asserts, "We do not need competencies for multicultural clients. We need methods and approaches that are effective with all kinds of clients" (p.72), then delineates aspects of such effective therapeutic relationships for universally effective counseling.

- Journal, essay and discuss with peers ideas for reaching out across cultural differences to make counseling known to be useful and valuable for all persons and cultures. Imagine as many cultural differences and situations/ contexts as possible, and generate as many methods for reaching out as possible.

- In order to deepen your understanding, journal and discuss as many scenarios as possible of ways in which counselors work with clients who have significant cultural differences. Remember to think broadly. For each scenario, imagine how the core conditions of counseling may help the persons establish a working and healing connection.

- Immerse yourself in the study of other cultures. It is often only minimally helpful to attempt to research a culture from which you appear to have a single or small number of clients. Generalizations about groups are not usually helpful in understanding single individuals. However, if you make cultures other than your own a consistent part of your scholarship, you may continually open your mind to the infinite range of possibilities for humans and gain a perspective that helps you better understand the culture that you bring to each relationship you form.

- Immerse yourself in the study of the cultures that you see yourself as growing from/identifying with. Deeply consider the meanings and influences of those cultures in the relationships you form.

- Read works by Virginia Axline, Gary Landreth, and Louise Guerney (see reference list for examples) and contemplate the differences between the worlds of children and the worlds of adults. Consider, speculate, journal, discuss, and thus begin your process of learning how to reach across those differences to make therapeutic connections, how you might apply the skills of *The Heart of Counseling* with children, and ways that you might benefit through the exchange of differences.

- Journal and discuss with your peers to generate as many examples of potential significant cultural differences as possible.

- Revisit the Primary Skill Objectives for this chapter and see if you have mastered them to your satisfaction at this time. If not, seek additional readings, practices, and discussions in order to master them to your satisfaction.

- Travel. Experience what it is like to be a stranger in a strange land.

13

Growing Your Therapeutic Relationship Skills and the Core Conditions of Counseling

Just be sure when you step, step with care and great tact,
and remember that Life's a Great Balancing Act.
Oh the Places You'll Go!

DR. SEUSS

PRIMARY SKILL OBJECTIVES

- Begin to conceptualize and be able to speculate how developing your therapeutic relationship skills, especially the core conditions, already has and can continue to benefit you personally.

- Understand and be able to offer examples of how the paradigm of the core conditions can guide you through the complexities of multifaceted relationships with clients.

- Understand how your therapeutic relationship skills improve your consultation skills. Be able to give examples from each core condition for consulting important persons in your clients' lives and your workplace.

- Begin to understand some of the ways that skills from therapeutic relationships can be applied in classrooms. Be able to give an example from each core condition for one or more classroom situations.

- Understand how skills from therapeutic relationships can be taught to clients for their use in relationships with others. Be able to give examples.
- Begin to conceptualize the many applications you can make of your therapeutic relationship skills, especially the core conditions.

FOCUS ACTIVITY

Journal, discuss with your peer group, and/or essay your answers to the following questions:

- What have you learned in your study of *The Heart of Counseling?* Of course, that is so broad a question, it would take a book to answer. So, start your journaling and/or discussion with the first thoughts that come to mind. Then build on those thoughts and develop them through journaling and/or discussion until you can boil your learning down to the few principles that stand as shorthand for you for the big volumes of your learning.
- Just what are counseling and therapeutic relationships? Narrow and develop your answers here in the same ways you did for the previous question.
- How is it that counseling and therapeutic relationships work? Narrow and develop your answers here in the same ways you did with the preceding questions, but pay particular attention to how your clients or practice partners seem to have changed and just what it was about the therapeutic relationships that you provided or practiced with them that contributed to those changes. Just what are the mechanisms that have facilitated or that you understand and expect will facilitate change?
- How have you changed through your study? How do you want to continue to improve yourself, and how can both giving and receiving the core conditions in counseling and other relationships benefit you in these improvements?
- In what other ways and in what noncounseling relationships and situations would you like and/or can you imagine employing and sharing your skills of therapeutic relationships and the core conditions?

INTRODUCTION

In this chapter, we want you to begin to solidify what you have learned and consider where else your journey into developing your skills for therapeutic relationships, especially Rogers' core conditions (1957, 1961, 1980), may take you. Our work in these areas has taken us a great many places so far.

DEVELOPING YOURSELF THROUGH DEVELOPING YOUR SKILLS IN THERAPEUTIC RELATIONSHIPS

We are better persons for having put our hearts into developing our skills of therapeutic relationships, especially for growing the core conditions of counseling within ourselves. At first we put ourselves into this endeavor in order to be the best counselors we could be. Soon we learned that the work was making us better persons. That process is ongoing and ever increasing. As we strive to warmly and nonpossessively accept and prize each client, we find we are better able to accept each other, ourselves, and a great many others in this way. As we strive to connect with our clients on an emotional level, we find our empathy skills continually growing, and in this way, we are also able to connect with ourselves, each other, and a great many others. As we strive to be genuine in our counseling sessions, to be aware of our feelings and underlying thoughts, to understand the meaning of those feelings and thoughts, and to make reasonable (although sometimes so instantaneous as to be intuitive) decisions as to when to express our experiences, we find that we can't "turn off" this awareness and that decisions as to when and how to express ourselves come more and more naturally throughout all our relationships. We also find that this *way of being,* as Carl Rogers would put it, has helped each of us develop a strong sense of self. We are both better able to assert ourselves in times of conflict and are better able to do so in empathic, warmly accepting, respectful, and even loving ways. We find that we are well in many of our relationships, although far from perfect, and we are thankful for that.

We find that developing our skills in building therapeutic relationships brings reciprocal warmth and successes in a great many relationships. Earlier in our lives and works, we were much more the fighters, metaphoric "dragon slayers"—we saw ourselves as powerful child and client advocates. We felt that as long as we were "fighting for a good cause," our tactics were also automatically good. Looking back, we now see that we may have sometimes lacked empathy for and warm acceptance of our "dragons." We may have relied more heavily on charisma and the power of persuasion than on respectful listening and the power of communication. Now that we more consistently combine empathy and acceptance with our assertiveness, we find we are able to be even more influential and more content with our works and ourselves. We are more accepting of what we are able to accomplish and what we aren't, but in accepting this we find we are able to accomplish more. We attribute a great deal of this influence, effectiveness, and contentment with our work to the skills and the *way of being* we continue to develop for building therapeutic relationships with our clients.

Relationships are complex. We use the paradigm of the core conditions to guide us through complexities. We offer examples of complexities guided by the core conditions in the following pages.

THE QUESTION OF MULTIPLE OR DUAL RELATIONSHIPS WITH CLIENTS

The American Counseling Association's code of ethics guides counselors to avoid dual relationships, when possible (ACA, 1995). The ethical codes of other mental health–related professional associations concur in advising against such relationships. However, the complexity of life and need for personal judgment allow that it isn't always possible to avoid dual or at least multifaceted relationships with clients. In many settings, counselors may have the same persons as individual clients and as group clients. This is not a dual relationship in the strictest sense, but it is a different relationship—in individual sessions, the person has your nearly undivided attention, while in the group, your attention is spread across other group members and the process of interactions. This is a relationship difference that we have noticed requires adjustment, especially for some child clients.

In agency and college settings, counselors usually do not have contact with clients away from work. The rule of thumb is usually that clients may approach you, but you will not approach them in chance meetings away from work. The reason for this is that approaching a client in public may make it clear that the person is a client and thus breaches the standard of confidentiality of those work settings. Our way of handling situations when we meet clients in public and there is an uncertain moment of how to proceed is to make eye contact and smile, without initiating any more obvious contact than that. This is usually sufficient to let clients know, "Of course I know you and I am open to you, but I am allowing you your space and privacy."

Regarding where to go from that initial contact, we use the self-awareness we have developed and consider the core conditions and skills of our counseling relationships to guide our decisions in interacting. For example, if a client begins in a noncounseling setting to tell one of us about a serious concern, the off-work counselor of this client would give a genuine and naturally empathic response to shift the communication to a counseling session where the counselor can more carefully and effectively respond: "I'm concerned about what you are telling me. I'd like you to bring it up in our counseling sessions so that I can give my full attention." You might also suggest the client call to make or move up an appointment, if needed and possible. This assumes you've not just been told of a situation of imminent danger or abuse, to which you must respond right away to keep the person safe, perhaps through guiding him to another mental health professional who is working at that time. Fortunately, in our experience, such responses have rarely been necessary, as most of our clients understand that we are away from work and just smile and keep it short.

It sometimes happens that clients and counselors are romantically attracted to each other. This should not be surprising as "two people sitting, privately talking of intimate subjects" to many people is the definition of a good date. When we've been fleetingly attracted to clients, it has usually passed when refocusing our empathy to see the whole person, rather than some idealized, partial version. If that has not been enough, it has helped to realize the part of ourselves that drives this momentary, mistaken attraction. For example, sometimes a feeling of attraction has reminded Jeff of his desire to rescue endangered others, to be Don Quixote to Dulcennia. Realizing, "Oh, it's that old part of me again," has usually made the attraction dissipate. If that is not enough and the attraction is affecting the counselor's work, any counselor in a situation of extensive romantic attraction to a client must consult with a peer or supervisor, and then perhaps refer the client to anther counselor (if speaking the attraction aloud in consultation or supervision doesn't work to dissolve it). It may also be necessary to seek counseling in order to discern and attend to the internal basis for the work-inhibiting attraction.

Jeff remembers sometimes enjoying the momentary flattery of knowing that a client is attracted to him.

> It has helped me not get too caught up with the complement to remember that since the client has had only a counseling relationship with me, it is not really even a crush on me but an attraction to the "me as counselor" that provides that person genuine, deep empathy and UPR. Even if I were single and not guided by moral and ethical prohibitions against seeking romance with someone in such a vulnerable position, I could not sustain the "me as counselor" across much more than the period of time that I already provide this relationship in counseling sessions. Rather, I need relationships that are reciprocal, and I take care to seek such relationships throughout life. In this way, I am not tempted to mistakenly seek dual relationships with persons who make themselves vulnerable to me in counseling.

Still, within reasonable limits, our relationships with clients are complex and can be rewarding on numerous levels. Jeff was once surprised by a former client who had artwork in a busy art opening that Jeff attended with friends.

> She came and found me through the crowd. She pulled me close to her side so she could tell me about her works as she showed them to me. I didn't remember her name or much about her. I did remember that she'd been a client but not recently and not for more than a brief period. I did have a memory of her as a warm, honest, and open person. I was glad to see her and have her show me her art. So, I gave her and her art my attention for the few moments that she seemed to want. As she showed me her art, which I easily liked, I accepted and valued it, just as I had accepted and valued her in her sessions. Just as I did for her in her counseling sessions, I accepted her art warmly for what it seemed to express to me, rather than making some great pretense of understanding or interpreting it. We shared a nice warm moment together, and both of us were satisfied when I returned to the friends I had come with.

Jeff remembers another client asking him for personal information near the end of his time in counseling.

> We had worked together off and on for a couple of years. I had seen him develop strong confidence and develop a set of directions and ways of being that satisfied him. In what we knew would be one of our last meetings, he asked me, "I know that in the past you've said that you want our sessions to be focused on me and what I want for myself, but now I just really want to know some about you. I'm curious and I think I've earned an answer."
>
> I enjoyed the strength of his question and couldn't/wouldn't try to argue the clear truth of it. He told me he wanted to know things like what I did for fun, what I spent my time on away from work, what my life was like. So, I told him some about my wife and our marriage, our foster daughter and other shared works, the fact that we share writing and teaching. That was somehow not what he wanted, so he asked about recreation. I told him of jogging, biking, and trips to the beach. When this didn't seem to be enough, I tried to explain how my work was also my play (that I liked to write, teach, and counsel). In only a few moments of hearing my answers, he turned the topic back to himself. Considering the level of strength he had developed, I don't believe my answers to his questions affected him much at all. I have the impression that he was mildly disappointed in my answers. He respected my answers, but perhaps he'd pictured me differently. Maybe he imagined that away from work I drove, or maybe even built, fancy sports cars, and jet-setted around the world consulting kings, queens, and gurus. My feel for our relationship in that moment let me know that it was okay to let him know the simple but lovable truth of me.

Counselors in schools necessarily have many faceted relationships with clients. As school adults, counselors in schools will, of course, take a role in supporting school rules, especially as they provide for safety. As classroom presenters, counselors in schools will at times need to do more talking than listening. We have not found these different relationships to be a detriment to therapeutic relationships in personal counseling. Clients, especially children, learn the different structures of different relationships in different places and times rather easily, or if not, this may be part of the reason why they are in counseling. To be too quick to doubt their ability to discern and adapt seems to be overly doubtful of the persons and their possibilities. When we have experienced students struggling, perhaps behaviorally, with the difference, we've simply asserted the present structure with empathy and UPR. For example, when repeatedly interrupted by a client in a classroom presentation, we might tenderly respond, "Jack, you're thinking it's okay to talk over me right now. One of the things you can't do is talk over me while I'm talking to the whole class."

THERAPEUTIC RELATIONSHIP
SKILLS IN CONSULTATION

We urge you to be creative and take your therapeutic relationship skills, especially the core conditions, with you into all of your professional consultations. We offer a few of examples here.

With Teachers

For counselors who work in schools, teacher consultation is a regular, important part of the job. A most typical example might be one in which a teacher comes to you as counselor complaining or worried about a student's behavior. Your natural, initial reaction might be to offer to see the student in counseling and/or to suggest changes that the teacher might make in order to better manage the student's behavior. However, in doing so, you would have whisked responsibility away from the teacher and onto yourself without being asked.

A better and more time-efficient approach would be to stop and listen first, to listen with empathy that allows the teacher to fully express his experience of the student, to listen with warm acceptance, which might require you to be aware of and set aside the prick of annoyance you feel at being asked to feel pain and sit with yet another person in a difficult situation, after all the students and parents you have served in your day. Such effort has not always been easy for us. But it has paid off.

Listening with empathy, you can help the teacher find that after venting his frustration, he knows more of what to do than he realized. If you are going to participate in helping this student or make suggestions for the teacher, listening well first will certainly help you make your actions more thoughtful and useful. Further, having listened fully and felt with this teacher will tend to make your actions or suggestions more credible and acceptable to the teacher, when or if the time for actions or suggestions comes.

Listening with UPR assists and enables your empathy (and vice versa). Listening with UPR also helps you see this teacher as a person struggling with a situation that is difficult for him in that moment, rather than as "complaint 1001 for the day" or a teacher who "*should* know better and *should* be a perfect teacher for all such problem situations."

Listening with a genuine sense of self helps you be secondarily aware of your experiences while primarily meeting the teacher with empathy and UPR; lets you listen therapeutically, as long as reasonable and productive; and allows you to assert your point of view when it will be most effective. In consultation, as in counseling, you are not a listening machine. You are a real person. In consultation, there is time to listen with empathy and UPR, and time to assert what you know.

Because we are fine counselors but probably not great teachers of school children, humility has often come naturally to us in consulting with teachers. Jeff remembers one teacher whom he knew to be very talented coming to him and expressing his frustration with certain students in his class frequently leading the larger group into conflicts. Jeff listened therapeutically first. In this way he avoided prompting the teacher to disengage and lose his talents for the situation. After listening well, the teacher asked Jeff if he could come and teach some sort of conflict resolution or manners in the classroom to solve the problem. Jeff felt a flash of fear at the request and realized that the base of that fear was that he believed if this great and skilled teacher could not teach the behaviors he wanted in the classroom, then he (Jeff) could not either. He did offer to keep the goal in mind for future classroom presentations and to provide a small group counseling experience for some from the class, in which they would have naturally occurring opportunities to resolve conflicts with each other in the group and thus learn skills to augment those they were learning in the classroom. Jeff's interventions certainly were helpful. However, a key to their success was the empowerment that this teacher received in his time in consultation, which served to keep his formidable skills in play.

With Parents and Loved Ones of Clients

We maintain this same humility and sense of self with parents and loved ones of clients as often as possible. In listening with empathy and UPR, we are often reminded that we do not know their child or loved one better than they do. Rather, we know their child or loved one differently, and both our understandings can benefit each other. The parents or loved ones can help a counselor understand their troubled child or loved one better. And the counselor's unique perspective of the client through her therapeutic relationship can help her educate these loved ones in how they may be most helpful. Additionally, a well-educated counselor has significant universal knowledge and wisdom of how persons develop mental health that can benefit many parents or loved ones in their desire to help.

With Other Professionals from Related Fields

A friend and former student's work comes to mind on this topic. This young counselor needed to work closely with a psychiatrist who initially tended to try to tell her what to do in counseling, what was possible, and what was not. Her approach in talking with him was to listen to his perspective and try to understand it, to listen to the hopes and fears for clients that he seemed to imply in his assertions. She didn't quite reflect what she thought was implied as it was not clear and her intuition told her it might prompt defensive reactions, but she listened to him with deep empathy and warm, caring acceptance for him as a person.

While this professional counselor was young and just gaining experience, she was confident that she had studied and trained well and found ways of counseling that were both effective and fit deeply for her. So, after listening with deep respect, she asserted for how she worked and why. As she had earned the right to assert through listening deeply, this psychiatrist listened to her in turn. We doubt he agreed with all that was said early on. Yet he seemed to respect this young professional counselor, as she had respected and listened to him and because she seemed to have deeply considered and to strongly believe in what she was saying and doing.

Eventually, he took the time to listen see the success of her work. Through her strong therapeutic relationships, she helped some clients succeed that he'd been ready to give up on. In time, he came to seldom allow psychiatric medications for clients that fell into her areas of service until those clients had tried counseling with her first. They became a strong mental health team. We see that their work together had a strong start and shifted from likely initial conflict through her listening first with all the empathy and UPR she could muster for the situation.

THERAPEUTIC RELATIONSHIP
SKILLS IN JOB TASK NEGOTIATION

We have learned that therapeutic relationship skills, especially the core conditions, can also be used in job task negotiations. For example, counselors in schools are often asked to take on tasks that are not counseling related. Some of this is only fair, as many people have to pull the weight of some tasks that ought not be their job, yet need to be done. But sometimes the noncounseling tasks requested of counselors go too far. Some tasks, like taking on a rule enforcement role beyond that of all adults in schools, are antithetical to counseling effectiveness. Some tasks just take too much precious time away from counseling. Again, we find that a sense of self, gained from growing the core conditions in your way of being, asserting your point of view with empathy and warm acceptance of other views and persons can be most effective in negotiating job tasks. A story from Jeff's work may help illustrate this:

> In a school where there were many pressing works needed from the counselor (me) early in my first year, my principal told me to begin keeping all students' cumulative files in my office and to oversee and maintain the quality and correctness of the content of the files. This would have been a huge undertaking and one that would have occupied much of my time and energy—probably an inordinate amount as I am admittedly terrible at clerical tasks. I tend to worry over and overwork them.
>
> My first impulse, back then, was to dig in my heals, lock horns, and say, "No way am I taking on that task!" Fortunately, I had cultivated enough respect for my principal to think again before responding. I had already felt with him as we worked through some early school year crises and enjoyed some initial successes. So, instead of initiating a fight, I initiated conversation to further discuss why he made this request. As I listened conversationally but therapeutically, it prompted him to explain his reasons for asking me, such as his concern for confidentiality, his thought that the task needed someone who understood testing, and his fear about the task needing to be done conscientiously and carefully. Listening and really trying to understand

his perspective helped me connect with his concerns and respond with alternative suggestions that were more respectful than conflictful.

Perhaps most importantly, building my approach to the potential conflict on empathy and UPR kept me from saying, "No, I will not do it!" which would have been insubordinate and certainly have locked us in conflict. But my approach also kept us in open communication over the task. I never said, "No." Rather, I said, "I wish you wouldn't ask me that, because _____," and "Please consider this alternate solution, _____." I continued to listen and strive to really understand his perspective and request, but I also continued to assert for alternatives and compromises. In this case, a compromise was eventually found that I was very satisfied with. In other cases, I conceded to his requests and simply took on helpful tasks (It isn't worth the time and energy to discuss everything that could be debatable).

THERAPEUTIC RELATIONSHIP
SKILLS IN CLASSROOMS

We find many applications for therapeutic relationship skills in the classroom. We help teachers to see the benefits of a more child-centered approach in the classroom (one where feelings are acknowledged, limits are set consistently with empathy, and a child learns to rely on self and intrinsic control) as opposed to an adult centered approach to the classroom (one where children repress feelings, conform to the group, blame each other, and rely on adults for rules and controls). We consult with and work with teachers in a proactive, and respectful manner—also using the core conditions as a guide in relating to teachers and staff who are often under great stress in the job of schooling our children. We offer a couple of examples here.

Empathy Sandwich

We learned the concept of the "empathy sandwich" from Bill Nordling (Nordling & Cochran, 1999) and it has multiple applications, including in the classroom. Empathy sandwich is Bill's term for a part of the limit-setting procedure in the National Institute for Relationship Enhancement model of child-centered play therapy (Guerney, 1983, 2001; Nordling & Cochran, 1999). In that model, limits set on children's behavior are *always preceded and followed by* expressions of empathy. The empathy is what "holds it all together," like two slices of bread. Limits are much better listened to and accepted when "sandwiched" between two "slices" of empathy. We have shared this concept with some teachers and have known other teachers to naturally set classroom limits this way.

A hypothetical classroom example follows: Let's say that without realizing it is going to upset him in that moment, the teacher asks a troubled student, Johnny, to do a certain task. For unexpected reasons, Johnny finds the task embarrassing and scary. So he responds to his teacher's request by yelling at her, "It's not fair! All you do is pick on me!" If the teacher simply and correctly responds that he is not allowed to speak to her in that tone, she and Johnny will be locked in a time-consuming power struggle that will not be conducive to learning. If she takes just a minute more to notice Johnny as a person in that moment and to feel what he feels, she could respond something like, "Johnny, something about what I asked you to do has hurt your feelings and you see it as completely unfair. Still, one of the things you cannot do is yell at me like that." Because she has acknowledged Johnny as a person and his feelings and thoughts in that moment, this helps him calm down, and he is more likely to do as she asks, be ready to problem solve with her, or at least withdraw from conflict. In some cases, a child in Johnny's situation may continue to show disgruntlement and reluctance but sulk off to do the task. In a child-centered play therapy session, the counselor would also respond to this action and way of being with empathy, but in the classroom, sometimes that step can be skipped, as noting his consent to do the task in front of the group may prompt more embarrassment and reluctance. However, the teacher may decide later, when she has a moment alone with Johnny, to say, "I know it was hard to stay on task today when you were angry. You did it anyway, and I want you to know that I noticed. Thanks." This would "top the sandwich off," so to speak, and would likely motivate more compliant and cooperative behaviors from Johnny in the future. More importantly, however, Johnny would benefit and grow from experiencing a relationship with a teacher who actively shows care and concern but still expects respect and responsibility.

In any example like this, it is essential that the teacher's empathy be real, at least as often as possible. If she is just going through the verbal motions, the same statement can be heard as condescension, rather than caring. If she speaks with real empathy, then he can know that she cares for and accepts him as a person, even though he has reached a line in behavior she will not let him cross. That kind of teacher-to-student interaction can be particularly important and helpful to students who are already working in counseling on issues such as whether they expect to be liked and cared for by others, and therefore how they wish to act toward others.

In our view, the reasonably consistent setting of structural limits, such as not allowing Johnny to yell at her in class, is an outgrowth of being a genuine person in the classroom. While the teacher is a caring person, she has her limits and ways that she is not willing to be treated.

Project Special Friend

When Jeff was a counselor serving an elementary school, he worked with a teacher to initiate a project to help an extraordinarily behaviorally and emotionally troubled first-grade boy achieve acceptance, safety, and a learning environment with his classmates. Once again, let's call this boy "Johnny." Johnny's behavior modification programs and counseling services were extensive, but to augment these actions, we asked his classmates to take a role in helping him. Because Johnny was at risk for soon being removed from the regular classroom, the approach was taken to allow his classmates to view his behavioral and emotional problems *as a difficulty and problem that they could have a role in help-ing change.* With his presence and with his consent in advance, we helped them accept him and develop empathy for him in spite of his aggressive behavior by helping them see that each person struggles with learning and development problems at different times. We explained what our actions would be in response to Johnny's wise (behavioral) choices and poor (behavioral) choices, and what our goals for his development were. We gave them a genuine sense of power in asking them to befriend him when he made wise choices, and ignore him or ask for help if needed when he made poor choices. But perhaps most importantly, the classmates saw his teacher model deep empathy for the pain that was at the core of Johnny's outward aggressive, hurtful behavior. This teacher was able to show acceptance and prizing of Johnny in spite of his misbehavior, and to show the genuineness of being a person who would also enforce reasonably firm and clear limits for his behavior in the classroom. This helped the whole class grow in their ability to show empathy and acceptance for each other, while also know-ing that it is okay to set limits, problem-solve, offer help, or get help when some-one is angry or sad and acting out. "Project Special Friend" has been repeated (with modifications according to situation, age, and grade) to allow success and inclusion for many other students we have worked with. Students who struggle with emotional and behavioral difficulties are too often quickly excluded from the regular classroom without being given a chance to learn how it feels to *gen-uinely be wanted* as part of a group. Asking for help from the whole class and modeling a safe environment that promotes understanding and care for all students—even those who are struggling emotionally and behaviorally—is pos-sible. One of our schools' greatest natural resources is the natural capacity for empathy that *all children have* and that *will develop* in classrooms where the core conditions are modeled, developed, and experienced.

TEACHING CLIENTS AND OTHERS TO USE THE SKILLS OF THERAPEUTIC RELATIONSHIPS IN THEIR RELATIONSHIPS

We have also found applications for teaching skills of therapeutic relationships for use in noncounseling relationships. We offer a few of examples here.

B. G. Guerney (1977) developed The Relationship Enhancement model for teaching family members to listen therapeutically and experience deep empathy with each other. We adapted parts of that model to facilitate conflict resolution among children. Our guide to this approach (Cochran, Cochran, & Hatch, 2002) helps elementary school-age children reach naturally occurring solutions to their conflicts through complete communication and deep empa-thy. Through this approach, the children continue to own their conflict, rather than having it taken away by an adult or other mediator, and learn to have a different relationship with the one they are in conflict with by seeing the world through that person's eyes. We have found it time efficient and effective long term in schools, day camps, and other settings.

In filial therapy (Guerney, 1964), the counselor or other mental health professional helps the parent develop a therapeutic relationship with her young child, based on child-centered play therapy (Guerney, 1983, 2001; Nordling & Cochran, 1999). This approach has been found helpful across a wide variety of settings and situations (e.g., Glazer & Kottman, 1994; Ginsberg, 2002; Harris & Landreth, 1997; Johnson, Bruhn, Winek, Krepps, & Wiley, 1999; Tew, Landreth, Joiner, & Solt, 2002).

"Oh, the Places You'll Go"*

While we have had very significant professional success using our skills of therapeutic relationships, we have only begun to scratch the surface of the power that is possible. While we have been successful counselors, our journeys into how to provide the best, most efficient therapeutic relationships that we can, to make ourselves the best possible tools for our clients, may extend on and on. While we have known of or have been involved in numerous applications of key elements of therapeutic relationships in noncounseling settings and relationships, the boundaries to applications such as these seem limitless.

We wish we could know all the places you will go with this work. There is much to do, and we are excited at the prospects of the therapeutic relationships you will build and foster. We wish you the very best and great success in your works.

ACTIVITIES AND RESOURCES
FOR FURTHER STUDY

- Journal, discuss, and/or essay how developing your therapeutic relationship skills, especially the core conditions, already have and/or may change you for the better.

- Meet with groups of your peers now and in the future to brainstorm new ways to apply the skills of therapeutic relationships, especially the core conditions. Also brainstorm ways you can participate in each others' favorite projects. Then, participate in each others' projects over time. Counseling and related works can be hard enough alone, and counselors and related mental health professionals often work alone with clients, so take opportunities to work together and support each other.

- If you have studied this book with a group of peers, you have already begun your peer support and peer supervision groups. Make plans now

---

*Dr. Suess, 1990, p. 1.

that you can and will follow up on later to meet regularly to support and supervise each other into your futures.

- Review journals like the *Person-Centered Journal,* the journals of the American Counseling Association, and journals of related professional associations for ongoing and exciting examples of applications of skills of therapeutic relationships and the core conditions in and out of counseling relationships. As you find them, share them with your peers.

- Review and reconsider the concepts presented in Chapter 1 and how you may have changed since first contemplating those concepts. As you continue to develop as a counselor, revisit those concepts periodically to keep them in mind and to mark the strengthening of your knowing what you know.

- Remember that developing your therapeutic relationship skills may be and probably should be a lifelong journey. Take heart and enjoy the ride.

Skill Support Resource A

1. **Self-Actualization:** Each of our human experiences is unique and our interpretations more so. Yet, there seems to be a common ideal of maturity, and our environment seems to want this ideal to happen. *It is important to your therapeutic relationships for you to value this uniqueness and trust the flow toward maturity.*

2. **Blocks Come into One's Path toward Ideal Maturity:** *Your role in therapeutic relationships can be to open a path, to clear a way, yet not push or choose for your clients.*

3. **The Capacity for Awareness, Reason, to Question, and Choose:** These capacities seem innate to humans. A high level of intelligence, at least as we often think of it, doesn't even seem necessary. *The therapeutic relationships that you develop can facilitate, open, or renew these natural capacities.*

4. **Interpretation of Experience and Development of Self-Concept:** The process of interpretation, making meaning, and forming beliefs about self and others begins precognitively and

the beliefs can be entrenched, though at a low level of awareness, before being given serious, conscious consideration. *The therapeutic relationships that you form can provide your clients safe places, interactions, and experiences from which to become fully aware and reform core interpretations of experience.*

5. **Awareness of Existence and Choice, and Questions of Self-Worth:** Awareness of existence becomes awareness of choice. Questions of self-worth evolve in a progression, something like the following: "Who am I?" "Who do I want to be?" "Am I okay, likeable, lovable, worthwhile?" "Do others like, love, respect me?" "How do I compare? Am I better, stronger, faster, good enough?" These thoughts can be at the barely aware level, like a radio in the far background, but still effect us greatly. *As you invite clients into therapeutic relationships, you provide a safe place and set of interactions that bring thoughts of self and self-worth to the surface for full consideration. This work helps your clients reevaluate their answers to core questions of existence and helps them take responsibility for how they choose to define themselves.*

6. **Self-Responsibility Is Anxiety Provoking:** Awareness of self leads to awareness of choice, which leads to awareness of responsibility. Responsibility can be awe-striking, intimidating. *To help your clients become responsible, first help them become aware. Then, remember that much of your job may be to experience and accept your clients' anxiety over their* responsibility *and to be with them and help them through that anxiety.*

7. **Awareness of Aloneness:** Because we are responsible for our actions, in a sense we are alone. This is often anxiety producing and seems to prompt humans to yearn for relationships. It also sometimes prompts us humans to seek to give our responsibilities to others. *The therapeutic relationships that you provide are powerful connections that lead to more and deeper relationships for your clients.*

8. **Emotions Are Useful:** Just as we need our physical feelings, we need our emotional feelings. *Always respect and value emotions in your therapeutic relationships. Shared emotions are the golden road to connecting and healing.*

9. **Every Action Is a Choice of Destiny:** There are no insignificant choices or communications. *Awareness of this can help you know that as you attend to your clients' communications and actions, all their communications and actions are significant.*

10. **The Internal World:** Remember that each person's internal world is a huge part of that person's existence, awareness, feelings, and actions. *In your therapeutic relationships, your clients' internal world is your focus and this focus brings strength to your work.*

11. **Locus of Control and Evaluation, and Being:** *Your focus on each client's internal world helps your clients develop a healthy internal locus of control and evaluation, and being.*

Skill Support Resource B

DO'S AND DON'TS OF REFLECTIVE LISTENING AND EXPRESSING EMPATHY

Most Basic Level— Reflecting Content

With these behaviors, the counselor is communicating, "I understand what you are telling me." Even more basically, the counselor is communicating, "I am truly listening." [Note to readers: The Do's and Don'ts of this most basic level are addressed in Chapter 1. However, as you will see in explanation, the basic and deeper levels are fully integrated in action.]

Do:

__ Use your personal version of listening body language.

__ Reflect your perception of what the client says.

__ Make your reflections declarative statements when sure; tentative declarations when unsure.

__ Keep your reflections short whenever possible.

__ Focus your reflections on the client's main point, or the things communicated that seemed most important and most emotionally laden to that client.

__ Be prepared for and accept corrections.

__ Interrupt the client carefully to make reflections. Considerations for the counselor here are (1) Will an interruption to reflect help the client clarify communication, thoughts, feelings?

(2) How much communication can you reasonably reflect in a short paraphrase before becoming overwhelmed?

__ Allow the client to own most silences.

Don't:

__ Allow interruptions for reflections to set up a hierarchy where your communications would seem more important than the client's.

__ Ask questions, except in rare circumstances, or state reflections in a questioning tone.

A Deeper Level— Expressing Empathy

With these behaviors, the counselor is communicating truths like the following: "I understand your situation," "I sense what is important to you," "I'm striving to feel as much as I can with you, to feel as if I were you," and, ultimately, "Through experiencing with you, I'm coming to understand you."

Do:

__ Focus your attention primarily on client emotions.

__ Strive to feel with your client, to feel what your client feels.

__ Reflect feelings and underlying thoughts that your client states.

__ Reflect feelings and underlying thoughts that you perceive your client to imply.

__ State your empathy in declarative statements when reasonably sure.

__ When unsure, state your empathy as tentative guesses from your struggle to understand your client's feelings and thoughts.

__ Use reflections to restate client feelings and underlying thoughts more clearly and directly.

__ Use reflections to restate client feelings and underlying thoughts more precisely and concisely.

__ Be prepared for and accept corrections of your empathy.

Don't:

__ Respond from a hidden agenda of what you believe clients should realize.

__ Do most of the talking.

__ Make "me too" statements.

Skill Support Resource C

SAMPLE INITIAL SESSION

REPORT ITEMS

Counselor: _____

Date: _____

Client/Case Number: _____

   I. Identifying Information

  II. Presenting Problem/Concerns

 III. History of Problem/Previous Interventions

 IV. Reason for Coming to Counseling *Now*

  V. Alcohol/Drug Use &/or Medical Concerns

  VI. Related Family History/Information

 VII. Major Areas of Stress

VIII. Academic/Work Functioning

  IX. Social Resources

   X. Initial Impressions or Understandings of the Person and Concerns

  XI. Treatment Plans

Skill Support Resource D

We invite you to challenge yourself to a thorough and tough essay-writing project to help you in knowing yourself in important ways that prepare you for counseling across cultural differences. This essay can be a major part of discovering what your values and beliefs about human nature are, and the origins of your values and beliefs. To give your learning through this essay the energy and attention it deserves, it should be difficult and time consuming. You should work through many drafts and revisions, considering and reconsidering both your ideas and your writing. You should work with one or more partners to provide ongoing editing and review, to challenge each others' ideas, and to provide support for those ideas. This can benefit both your writing and your learning from others' ideas.

Limit yourself to seven pages to explain your beliefs, thoughts, feelings, and life experiences, addressing the following items: (1) your beliefs on how personalities are formed and how they change, (2) key notions of your beliefs about human nature, (3) the main values you live by, how those values became your values, and how those central values might influence you as a counselor (meaning what kinds of helpful things those values might naturally prompt you to do), (4) what gives you a sense of meaning and purpose in life and how this meaning seems to be related to your desire to help others, (5) your view of counseling at this time, related to and formed by these beliefs, notions, and values; as well as how you might briefly articulate your view of counseling to a client or others in a client's life, and (6) what you believe typical and useful counseling goals would be and what qualities generally make for an excellent counselor, based on your answers to items 1–4. Illustrate or support every assertion you make with where or how in your background of experiences you came to hold that view or belief. None of your views, beliefs, or values simply came to you from the sky or by some other accident. Each of them developed from your experiences and the meanings you made of experiences. Additionally, when you find contradictions in your views, values, and beliefs, work to resolve them. Finding contradictions can make for important moments in learning to know yourself.

We suggest the short format so that you will force yourself to think very carefully instead of submitting to the temptation to simply list and support your views, values, and beliefs. For example, if we were to address any one of the items of this essay, we could go on for numerous pages listing and supporting aspects of ourselves. Actually, that would be a good way to start your thought process for this essay. Then, once you have a long and exhaustive list, search to see which items fall into groups or categories. Then, discern which of the groups and categories are most important to you and why, so that you include your top priorities, most important views, values, and beliefs.

We also suggest the short format because when counselors have moments to express their beliefs to clients or important persons in clients' lives, those moments usually pass very quickly. So, we'd like you to be ready.

Be creative in discerning how you can include the ideas that are really important to you within the limited space allowed. For example, could you address more than one of the items together and still be clear? Might you tell one or a few examples that pull together and support a number of your ideas? Don't limit yourself to the first idea for formatting the essay that occurs to you.

References

American Counseling Association. (1995). *Code of ethics and standards of practice.* Alexandria, VA: Author.

The American Heritage Dictionary (2nd ed.). (1982). Boston, MA: Houghton Mifflin.

Augsburger, D. W. (1986). *Pastoral counseling across cultures.* Philadelphia, PA: Westminster Press.

Axline, V. M. (1947). *Play therapy: The inner dynamics of childhood.* Cambridge, MA: The Riverside Press.

Beck, A. T., & Weishaar, M. E. (1989). Cognitive therapy. In R. J. Corsini and D. Wedding (Eds.), *Current psychotherapies* (4th ed.) (pp. 285–322). Itasca, IL: F. E. Peacock.

Bergin, A. E., & Lambert, M. J. (1978). The evaluation of therapeutic outcomes. In S. L. Garfield and A. E. Bergin (Eds.), *Handbook of psychotherapy and behavior change* (2nd ed.) (pp. 139–189). New York: John Wiley.

Blake, C. R. L., & Garner, P. (2000). *"We may give advice but we can never prompt behavior": Lessons from Britain in teaching students whose behavior causes concern.* (ERIC Document Reproduction Service No. 442209).

Bohart, A. C., Elliot, R., Greenberg, L. S., & Watson, J. C. (2002). Empathy. In J. C. Norcross (Ed.), *Psychotherapy relationships that work* (pp. 89–108). New York: Oxford University Press.

Bourne, E. J (2000). *The anxiety and phobia workbook.* Oakland, CA: New Harbinger.

Bozarth, J. D. (1998). *Person-centered therapy: A revolutionary paradigm.* Ross-on-Wye, UK: PCCS Books.

Bozarth, J. D., & Wilkins, P. (2001). *Unconditional positive regard.* Ross-on-Wye, UK: PCCS Books.

Brooks, R., & Goldstein, S. (2001). *Raising resilient children: Fostering strength, hope, and optimism in your child.* (ERIC Document Reproduction Service No. ED 449921).

Chiu, E. (1998). A patient who changed my practice. *International Journal of Psychiatry in Clinical Practice, 2,* 231–232.

Clifford, E. F. (1999). A descriptive study of mentor–protégé relationships, mentors' emotional empathic tendency, and protégés' teacher self-efficacy belief. *Early Child Development and Care, 156,* 143–154.

Cochran, J. L. (1996). *The status of services to students with conduct disorder by their elementary school counselors.* Dissertation Abstracts International, 57, 03A (University Microfilms No. AAI96-24171).

Cochran, J. L., & Cochran, N. H. (1999). Using the counseling relationship to facilitate change in students with conduct disorder. *Professional School Counseling, 2,* 395–403.

Cochran, J. L., Cochran, N. H., & Hatch, E. J. (2002). Empathic communication for conflict resolution among children. *Person-Centered Journal, 9,* 101–112.

Cohen, J. (1994). Empathy toward client perception of therapist intent: Evaluating one's person-centeredness. *Person-Centered Journal, 1,* 4–10.

Corey, G. (1996). *Theory and practice of counseling and psychotherapy* (5th ed.). Pacific Grove, CA: Brooks/Cole.

Corey, G., Corey, M. S., & Callanan, P. (1998). *Issues and ethics in the helping professions* (5th ed). Pacific Grove, CA: Brooks/Cole.

Cramer, D. (1990). Towards assessing the therapeutic value of Rogers' core conditions. *Counselling Psychology Quarterly, 3,* 57–66.

Cramer, D. (1994). Self-esteem and Rogers' core conditions in close friends: A latent variable path analysis of panel data. *Counselling Psychology Quarterly, 7,* 327–337.

Demanchick, S. P., Cochran, N. H., and Cochran, J. L. (2003). Person-centered play therapy for adults with developmental disabilities. *International Journal of Play Therapy, 12,* 47–65.

Douglas, C. (2005). Analytical psychotherapy. In R. J. Corsini & D Wedding (Eds.), *Current psychotherapies* (7th ed.). Belmont, CA: Brooks/Cole.

Duan, C., Rose, T. B., & Kraatz, R. A. (2002). Empathy. In G. S. Tryon (Ed.), *Counseling based on process research:*

Applying what we know (pp. 197–231). Boston, MA: Allyn & Bacon.

Ellis, A. (1989). Rational-emotive therapy. In R. J. Corsini & D. Wedding (Eds.), *Current psychotherapies* (4th ed.) (pp. 197–240). Itasca, IL: F. E. Peacock.

Ellis, A. (2005). Rational-emotive behavior therapy. In R. J. Corsini & D Wedding (Eds.), *Current psychotherapies* (7th ed.). Belmont, CA: Brooks/Cole.

Ellis, A., & Dryden, W. (1997). *The practice of rational-emotive behavior therapy.* New York: Springer.

Ellis, A., & Harper, R. A. (1997). *A guide to rational living.* North Hollywood, CA: Wilshire Books.

Everding, H. E., & Huffaker, L. A. (1998). Educating adults for empathy: Implications of cognitive role-taking and identity formation. *Religious Education, 93,* 413–430.

Farber, B. A., & Lane, J. S. (2002). Positive regard. In J. C. Norcross (Ed.), *Psychotherapy relationships that work* (pp. 175–194). New York: Oxford University Press.

Ginsberg, B. G. (2002). The power of filial relationship enhancement therapy as an intervention in child abuse and neglect. *International Journal of Play Therapy, 11,* 65–78.

Glazer, H. R., & Kottman, T. (1994). Filial therapy: Rebuilding the relationship between parents and children of divorce. *Journal of Humanistic Counseling Education and Development, 33,* 4–12.

Grafanaki, S. (2001). What research has taught us about congruence. In G. Wyatt (Ed.), *Rogers' therapeutic conditions: Evolution, theory and practice. Volume 1: Congruence* (pp. 18–35).

Grafanaki, S., & McLeod, J. (1995). Client and counselor narrative accounts of congruence during the most helpful and hindering events of an initial counseling session. *Counselling Psychology Quarterly, 8,* 311–324.

Graves, J., and Robinson, J. (1976). Proxemic behavior as a function of inconsistent verbal and nonverbal messages. *Journal of Counseling Psychology, 23,* 333–338.

Grove, P. B. (Ed.). (1966). *Webster's third new international dictionary*. Springfield, MA: G. & C. Merriam.

Guerney, B. G., Jr. (1964). Filial therapy: Description and rationale. *Journal of Consulting Psychology, 28,* 304–360.

Guerney, B. G., Jr. (1977). *Relationship enhancement.* San Francisco, CA: Jossey-Bass.

Guerney, B. G., Jr. (2002, September 30). Deep empathy, part II. *Relationship Enhancement Newsletter, 4,* 2–3.

Guerney, L. F. (1976). Filial therapy program. In H. L. Benson, (Ed.), *Treating relationships* (pp. 67–91). Lake Mills, IA: Graphic Publishing.

Guerney, L. F. (1983). Client-centered (nondirective) play therapy. In C. E. Schaefer, & K. J. O'Connor (Eds.), *Handbook of play therapy* (pp. 21–64). New York: John Wiley & Sons.

Guerney, L. F. (1995). *Parenting: A skills training manual* (5th ed.). North Bethesda, MD: Institute for the Development of Emotional and Life Skills.

Guerney, L. F. (2001). Child-centered play therapy. *International Journal of Play Therapy, 10,* 13–31.

Haaga, D. A. F., Rabois, D., & Brody, C. (1999). Cognitive behavior therapy. In M. Hersen, & A. Bellack (Eds.), *Handbook of comparative interventions for adult disorders* (2nd ed.) (pp 48–61). New York: John Wiley & Sons.

Haase, R., & Tepper, D. (1972). Nonverbal components of empathic communication. *Journal of Counseling Psychology, 19,* 417–424.

Harris, Z. L., & Landreth, G. L. (1997). Filial therapy with incarcerated mothers: A five week model. *International Journal of Play Therapy, 2,* 53–73.

Havens, L. L. (1986). *Making contact: Uses of language in psychotherapy.* Cambridge, MA: Harvard University Press.

Jacobs, M. (1988). *Psychodynamic counselling in action.* London: Sage.

Johnson, L., Bruhn, R., Winek, J., Krepps, J., & Wiley, K. (1999). The use of child-centered play therapy and filial therapy with Head Start families: A brief report. *Journal of Marital and Family Therapy, 25,* 169–176.

Jung, C. G. (1935/1956). Two essays of analytical psychology. Collected works, Vol. 17. In C. G. Jung, *The collected works of C. G. Jung* (22 Volumes). Princeton, NJ: Princeton University Press.

Keijsers, G. P. J., Schaap, C. P. D. R., & Hoogduin, C. A. L. (2000). The impact of interpersonal patient and therapist behavior on outcome in cognitive-behavioral therapy: A review of empirical studies. *Behavior Modification, 24,* 264–297.

Kinast, R. L. (1984). The dialogue decalogue: A pastoral commentary. *Journal of Ecumenical Studies, 21,* 311–318.

Kohut, H. (1984). *How does analysis cure?* Chicago, IL: University of Chicago Press.

Kountz, C. (1998, April). *The anxiety of influence and the influence of anxiety.* Paper presented at the Annual Meeting of the Conference on College Composition and Communication, Chicago, IL. (ERIC Document Reproduction Service N. ED448461)

Krumboltz, J. D., Becker-Haven, J. F., & Burnett, K. F. (1979). Counseling psychology. *Annual Review of Psychology, 30,* 555–602.

Lambert, M. J., & Okiishi, J. C. (1997). The effects of the individual psychotherapist and implications for future research. *Clinical Psychology: Science and Practice, 4,* 66–75.

Landreth, G. L. (2002). *Play therapy: The art of the relationship.* New York: Brunner-Routledge.

Lenaghan, A. (2000). Reflections on multicultural curriculum. *Multicultural Education, 7,* 33–36.

LeShan, L. (1974). *How to meditate: A guide to self-discovery.* Boston, MA: Little, Brown and Company.

Lickona, T. (2000). Sticks and stones may break my bones AND names WILL hurt me: Thirteen ways to prevent peer cruelty. *Our Children, 26,* 12–14.

Lietaer, G. (1984). Unconditional positive regard: A controversial basic attitude in client-centered therapy. In R. Levant & J. Shlien (Eds.), *Client-centered therapy and the person-centered approach: New directions in theory, research, and practice* (pp. 41–58). New York: Praeger.

Lietaer, G. (1993). Authenticity, congruence and transparency. In D. Brazier (Ed.), *Beyond Carl Rogers* (pp. 17–46). London: Constable.

Lineham, M. M. (1997). Validation and psychotherapy. In A. C. Bohart & L. S. Greenberg (Eds.), *Empathy reconsidered: New directions in psychology* (pp. 353–392). Washington, DC: American Psychological Association.

Lo Bianco, J. (1999). *Training teachers of language and culture: Language Australia research, policy and practice papers.* Melbourne (Victoria), Australia. (ERIC Document Reproduction Service No. ED 449 675).

Lockhart, W. (1984). Rogers' necessary and sufficient conditions' revisited. *British Journal of Guidance and Counselling, 12,* 112–123.

Luborsky, L., Crits-Christoph, P., McLellan, T., Woody, G., Piper, W., Liberman, B., Imber, S., & Pilkonis, P. (1986). Do therapists vary much in their success? Findings from four outcome studies. *American Journal of Orthopsychiatry, 51,* 501–512.

May, R. (1989). *The art of counseling.* New York: Gardner.

McCarthy, M. (1992). Empathy: A bridge between. *Journal of Pastoral Care, 46,* 119–128.

Mearns, D., & Thorne, B. (1988). *Person-centred counselling in action.* London: Sage.

Myers, S. (2000). Empathic listening: Reports on the experience of being heard. *Journal of Humanistic Psychology, 40,* 148–173.

Nordling, W. (1998, April). *Child-centered play therapy.* Workshop series sponsored by the National Institute for Relationship Enhancement (NIRE), Baltimore, MD.

Nordling, W., & Cochran, N. H. (1999, April). *Child-centered play therapy.* Workshop series sponsored by the National Institute for Relationship Enhancement (NIRE), Savannah, GA.

O'Hara, M., & Jordan, J. V. (1997). Relational empathy: Beyond modernist egocentrism to post modern holistic contextualism. In A. C. Bohart & L. S. Greenberg (Eds.), *Empathy reconsidered: New directions in psychology* (pp. 295–320). Washington, DC: American Psychological Association.

Orlinsky, D. E., & Howard, K. I. (1978). The relation of process to outcome in psychotherapy. In S. L. Garfield and A. E. Bergin (Eds.), *Handbook of psychotherapy and behavior change* (2nd ed.) (pp. 283–330). New York: John Wiley.

Orlinsky, D. E., & Howard, K. I. (1986). Process and outcome in psychotherapy. In S. L. Garfield and A. E. Bergin (Eds.), *Handbook of psychotherapy and behavior change* (3rd ed.) (pp. 331–381). New York: John Wiley.

Patterson, C. H. (1984). Empathy, warmth and genuineness: A review of reviews. *Psychotherapy, 21,* 431–438.

Peacock, S. (1999). Internal mental space. *Therapeutic Communities: International Journal for Therapeutic and Supportive Organizations, 20,* 301–314.

Perls, F. S. (1970). Four lectures. In J. Fagan and I. L. Shepherd, (Eds.), *Gestalt therapy now* (pp. 14–38). Palo Alto, CA: Science and Behavior Books.

Peschken, W. E., & Johnson, M. E. (1997). Therapist and client trust in the therapeutic relationship. *Psychotherapy Research, 7,* 439–447.

Purton, C. (1996). The deep structure of the core conditions: A Buddhist perspective. In R. Hutterer, G. Pawlowsky, P. F. Schmid, & R. Stipsits (Eds.), *Client-centered and experiential psychotherapy: A paradigm in motion* (pp. 455–467), Frankfurt-am-Main: Peter Lang.

Rogers, C. R. (1957). The necessary of sufficient conditions of therapeutic personality change. *Journal of Consulting Psychology, 21,* 95–103.

Rogers, C. R. (1961). *On becoming a person.* Boston, MA: Houghton Mifflin.

Rogers, C. R. (1980). *A way of being.* Boston, MA: Houghton Mifflin.

Rogers, C. R. (1987). Rogers, Kohut, and Erickson: A personal perspective on some similarities and differences. In J. K. Zeig (Ed.), *The evolution of psychotherapy* (pp. 179–187). New York: Brunner/Mazel.

Satir, V. (1987). The therapist story. In M. Baldwin & V. Satir (Eds.), *The use of self in therapy* (pp. 17–25). New York: Haworth Press.

Scharf, R. S. (1996). *Theories of psychotherapy and counseling: Concepts and cases.* Pacific Grove, CA: Brooks/Cole.

Scheffler, T. S., & Naus, P. J. (1999). The relationship between fatherly affirmation and a woman's self-esteem, fear of intimacy, comfort with womanhood and comfort with sexuality. *Canadian Journal of Human Sexuality, 8,* 39–45.

Schuster, R. (1979). Empathy and mindfulness. *Journal of Humanistic Psychology, 19,* 71–77.

Seuss, Dr. (1990). *Oh, the places you'll go!* (Pop-Up Version). New York: Random House Children's Books (p. 9).

Sherer, M., & Rogers, R. (1980). Effects of therapist's nonverbal communication of rated skill and effectiveness. *Journal of Clinical Psychology, 36,* 696–700.

Shostrom, E. L. (Producer/Director). (1965). *Three approaches to psychotherapy.* Corona del Mar, CA: Psychological and Educational Films.

Sweeney, J., & Whitworth, J. (2000). *Addressing teacher supply and demand by increasing the success of first-year teachers.* (ERIC Document Reproduction Service No. ED 440 944).

Tew, K., Landreth, G. L., Joiner, K. D., & Solt, M. D. (2002). Filial therapy with parents of chronically ill children. *International Journal of Play Therapy, 11,* 79–100.

Truax, C., & Carkhuff, R. (1967). *Toward effective counseling and psychotherapy.* Chicago: Aldine.

Truax, C., & Mitchell, K. (1971). Research on certain therapist interpersonal skills in relation to process and outcome. In A. Bergin & S. Gardfield, (Eds.), *Handbook of psychotherapy and behavior change* (pp. 299–344). New York: John Wiley and Sons.

Tudor, K., & Worrall, M. (1994). Congruence reconsidered. *British Journal of Guidance and Counselling, 22,* 197–206.

Tyson, J., & Wall, S. (1983). Effect of inconsistency between counselor verbal and non-verbal behavior on perception of counselor attributes. *Journal of Counseling Psychology, 30,* 433–437.

van Ryn, M., & Heaney, C. A. (1997). Developing effective helping relationships in health education practice. *Health Education and Behavior, 24,* 683–702.

Wampold, B. E. (2001). *The great psychotherapy debate: Models, methods, and findings.* Mahwah, NJ: Erlbaum.

Watson, N. (1984). The empirical status of Rogers' hypotheses of the necessary and sufficient conditions for effective psychotherapy. In R. F. Levant and J. M. Shlien (Eds.), *Client-centered therapy and the person-centered approach: New directions in theory, research, and practice* (pp. 17–40). New York: Praeger.

Wilkins, P. (2000). Unconditional positive regard reconsidered. *British Journal of Guidance and Counselling, 28,* 23–36.

Wingert, P. (1999, March 22). The edge of kindness: A veteran kindergarten teacher on what she's learned. *Newsweek,* p. 78.

Wubbolding, R. E. (1996). Working with suicidal clients. In B. Herlihy & G. Corey (Eds.), *ACA ethical standards casebook* (5th ed.). Alexandria, VA: American Counseling Association.

Wyatt, G. (2001). *Rogers' therapeutic conditions. Volume 1: Congruence.* Ross-on-Wye, UK: PCCS Books.

SPECIAL REFERENCE LIST
FOR PRE-CHAPTER QUOTES

Introduction

Saint-Exupery, A. (1943). *The little prince.* Translated by Richard Howard. San Diego, CA: Harcourt, 2000 (p. 63).

Muhammad. As quoted in Fadiman, J., & Frager, R. (1998). *Essential Sufism.* Edison, NJ: Castle Books (p. 89).

Chapter 1

Fadiman, J., & Frager, R. (1998). *Essential Sufism.* Edison, NJ: Castle Books (p. 101).

Chapter 2

Barrett-Lennard, G. T. (1988). Listening. *Person-Centered Review, 3,* 410–425. (p. 410).

Wayne, J. (1907–1979). As quoted in E. M. Beck (Ed.), J. Bartlett (compiler), *Familiar quotations* (15th ed.). (1980). Boston, MA: Little, Brown and Company (p. 870).

Armstrong, L. (1900–1971). As quoted in E. M. Beck (Ed.), J. Bartlett (compiler), *Familiar quotations* (15th ed.). (1980). Boston, MA: Little, Brown and Company (p. 847).

Chapter 3

Jung, C. G. (1953). Psychological reflections: A Jung anthology: Collected works, 9. In J. Jacobi (Ed.) *Psychological aspects of the modern archetype,* (p. 32). As quoted in *Bartletts's familiar quotations* (15th ed.). (1980). Boston, MA: Little Brown and Company (p. 753).

Fadiman, J., & Frager, R. (1998). *Essential Sufism.* Edison, NJ: Castle Books (p. 101).

Roethke, T. (1969). The waking. In *The contemporary American poets.* New York: The New American Library (p. 234).

Chapter 4

T. S. Eliot. Philip Massinger. In *Bartlett's familiar quotations* (15th ed.). (1980). Boston, MA: Little Brown and Company (p. 809).

Chapter 5

Lama, The Dalai. (2002). *The Dalai Lama's little book of wisdom.* New York: Barnes and Noble Books (p. 103).

Niebuhr, Reinhold. The Irony of American History. As quoted in E. M. Beck (Ed.), J. Bartlett (compiler), *Familiar quotations* (15th ed.). (1980). Boston, MA: Little, Brown and Company (p. 823).

Chapter 6

Lindbergh, A. M. (1955). Gift from the sea, Chapter 2. As quoted in E. M. Beck (Ed.), J. Bartlett (compiler), *Familiar quotations* (15th ed.). (1980). Boston, MA: Little, Brown and Company (p. 867).

Blake, William. The marriage of heaven and hell [1790–1793] Proverbs of Hell, l. 69. As quoted in E. M. Beck (Ed.), J. Bartlett (compiler), *Familiar quotations* (15th ed.). (1980). Boston, MA: Little, Brown and Company (p. 404).

Wampold, B. E. (2001). *The great psychotherapy debate: Models, methods, and findings.* Mahwah, NJ: Erlbaum.

Chapter 7

Aristotle (quoting a Proverb) in Nicomachean Ethics, bk. V, ch. 4. As quoted in E. M. Beck (Ed.), J. Bartlett (compiler), *Familiar quotations* (15th ed.). (1980). Boston, MA: Little, Brown and Company (p. 88).

Chapter 8

Ming-Dao, Deng. (1992). *Daily meditations.* New York: Harper Collins (p. 128).

Chapter 9

Da Vinci, L. *The notebooks* [1508–1518]. Translated by Edward MacCurdy, vol. I, ch.2. As quoted in E. M. Beck (Ed.), J. Bartlett (compiler), *Familiar quotations* (15th ed.). (1980). Boston, MA: Little, Brown and Company (p. 152).

Chapter 10

Alighieri, Dante. (1265–1321). *The Divine Comedy,* "The Inferno," canto 1, line 1, translated by John D. Sinclair. As quoted in *Knowledge cards.* Rohnert Park, CA: Pomegranate Publications.

Hayden, R. (1996). Those winter Sundays. In A. Poulin Jr. and M. Waters (Eds.), *Contemporary American Poetry* (p. 216). New York: Houghton Mifflin.

Chapter 11

Ming-Dao, Deng. (1992). *Daily meditations.* New York: Harper Collins (p. 98).

Chapter 12

Mead, M. (1935). Sex and temperament in three primitive societies, concluding remarks. As quoted in E. M. Beck (Ed.), J. Bartlett (compiler), *Familiar quotations* (15th ed.). (1980). Boston, MA: Little, Brown and Company (p. 853).

Chapter 13

Seuss, Dr. (1990). *Oh, the places you'll go!* (Pop-Up Version). New York: Random House Children's Books (p. 9).

Index

A

academic functioning, 134
acceptance, 82, 84, 88
  of client's pace, 169
  healing of, 173
  initial, 173
  lack of, 171
  in managing crises, 187
actions
  as choices of destiny, 15
  therapist's, as communication, 25
adolescents
  communicating UPR to, 92
  controlled by others, 156
  deciding to continue each week, 232
  explaining reasons for referral to, 143
  logistics of sessions with, 151
  planning for ongoing counseling
    with, 127
  self-definition of, 12
adults
  communicating UPR to, 92
  controlled by others, 156
  deciding to continue each week, 232
  explaining reasons for referral to, 143
  logistics of sessions with, 151

  planning for ongoing counseling
    with, 127
  self-definition of, 12
agenda, 94
  as barrier to communicating UPR, 92
alcohol use, 133
aloneness, awareness of, 13
American Counseling Association, code of
    ethics, 269
anger, and progress, 230
anxiety
  and family history, 133
  history of, 132
  of new client, 141
  and progress, 229
  and reasons for coming to counseling
    now, 132
  social resources for, 135
  therapeutic listening and, 37
  treatment plans for, 138
  understanding situation, 136
awareness, capacity for, 9

B

being, value of, 16
belligerence, 231

body language, 26, 27, 28, 62, 66
burnout
  avoiding, 94
  as barrier to UPR, 93

C

career indecision
  and family history, 134
  history of, 132
  reasons for coming to counseling
      now, 132
  social resources for, 135
  treatment plans for, 138
  understanding situation, 136
case examples, general explanation of, 3
case notes, ongoing, 140
challenging, 23
change, defined, 228
child-centered play therapy, 15, 83,
      275, 278
children, 275
  in case examples, 4
  communicating UPR to, 92
  in cross-cultural therapeutic
      relationships, 258
  domestic violence involving, 213
  egocentrism of, 11
  ending therapeutic relationship with,
      226
  genuineness with, 122
  parents of, explaining counseling to, 164
  planning for ongoing counseling with,
      127
  reasonable goals for, 146
  self-definition of, 12
  UPR for, 81
choice, 12
  awareness of, 13
  of destiny, 15
choice moments, 15
choosing, 9
clients
  allowing room for, 176
  asking for counselor's experience of
      them, 119
  asking for guidance, 156
  with behaviors counter to counselor's
      moral constructs, 96
  communication of, big picture of, 176
  crises of, 183
  with difficulty starting therapeutic
      relationship, 180
  explaining counseling to, 165
  expression by, 121

future orientation of, 193
getting started with, 124–149
having great discomfort with silence, 159
how therapist sees, 24
initial discomfort of, 174
initial impressions of, 139
initially reluctant, 227
insisting on quick solutions, 157
level of emotions of, 173
logistics of sessions with, 151
with lowered impulse control, 193
with lowered inhibitions, 193
multiple or dual relationships with, 269
needing help in getting started, 167–182
needs of, other than counseling, 128
new. See new client
not seeing value of counseling, 259
pace of, 174, 176
personal connections with, 121
persons significant to, 164
quiet, 170
reasons for coming to counseling at
      present time, 132
reluctance to end therapeutic
      relationship, 244
at risk of harming others, 206
starting to use therapeutic relationship,
      170
starting where client is, 173
stories of, 40
sudden change in way of being, 193
talkative, 170
understanding how counseling may
      work for them, 155
understanding how to use counseling,
      162
understanding of, 139
urge to fix immediately, 39
wanting a quick fix, 40
who are difficult to like, 96, 115
who attend sporadically, 160
who don't know where to start, 161
Cochran, Jeff, 2
Cochran, Nancy, 2
cognitive behavioral therapy, 49
communication
  from clients, broad view of, 24
  empathy in, 53
  of genuineness, 110
  important, 41
  of UPR, 90
conditional positive regard, 84
conditions loop, 107
conduct-disordered behaviors, 107, 122
  empathy and, 52

confidentiality
  breaking of, with suicidal client, 196
  explaining to new client, 142
confronting, 23
confusion, useful, 129
congruence, 106, 108
connection, empathy in, 53
control
  letting go of, 56
  locus of, 16
core concepts, 49, 260
  summary of, 287–288
core conditions, 106, 108, 139
  and genuineness, 107
  growing, 266–279
core questions, 12
core self, emotions as connection to, 50
corrections, acceptance of, 91
counseling
  actions that help clients struggling in
      starting, 173
  concepts of, 5, 6
  eleven concepts of, 5–18
  explaining how it works, 155
  explaining reasons for, 179
  explaining to persons significant to
      client, 164
  genuineness in, 104
  interaction in, structure of, 155
  problems in explaining use of, 165
  single session of, 127
  starting, 167, 168
counseling sessions
  ending of, 151
  ending on time, exceptions to, 154
  endings of, letting clients own, 152
  initial, 126
  length of, 151
  time warnings for, 152, 154
  writing notes during, 130
counseling-related assessment, 129
counselors
  analytic minds of, 95
  avoiding limiting clients' expression, 121
  beginning, 165, 214
  being who they are, 113
  client's feelings about, 38
  competency of, 217
  consulting with other professionals, 217
  culture of, 259
  experiences of clients, 110
  expressing negative experiences of
      clients, 117
  expressing positive experiences of
      clients, 117

fulfilling own needs through clients, 95
having multiple or dual relationships
    with clients, 269
inhibiting client's beginning use of
    counseling, 169
initial discomfort of, 174
judging thoughts of, 97
maturity of, 115
messages of, 106
misattributed responsibility of, 55
not wanting to let go of client, 243
panic of, when dealing with dangerous
    situations, 214
reactions of, self-awareness of, 116
responsibility of, in hospitalization of
    client, 202
self-acceptance of, 116
self-awareness of, 116
self-confidence of, 217
self-development of, 115
self-expressions of, 114
self-honesty of, 116
shoulds of, self-awareness of, 116
stating feelings, 188
stating reactions when they interfere
    with empathy and UPR, 111
trying too hard, 169
who "know better" than clients, 92
worrying about motivating clients, 169
crises
  assessment of, 191
  empathy in, 215
  erring on side of caution in, 189
  explaining assessment and decision
      process in handling, 187
  feelings in, 215
  focus activity for, 184
  management of, 183
  planning for, 189, 191
  principles of managing, 186
  reflections during, 188
  stating feelings during, 188
  therapeutic relationship skills in, 185
crisis management panic mode, 218
cross-cultural therapeutic relationships,
    250–265, 292–293
  assumptions in, 255
  becoming accessible in, 263
  beginning counselors and, 261
  connecting in, 262
  difference as advantage to counselor,
      261
  discomfort involved in, 252
  focus activity for, 251
  humility in, 255

cross–cultural therapeutic relationships
  (*continued*)
  involving children, 258
  missing content for cultural context,
    261
  missing feelings for cultural context, 261
  opportunity to experience diversity of
    clients, 261
  reaching out in, 263
  sensing need for context education in,
    262
  sensing need for information in, 262
  significant value differences in, 262
  thinking broadly in, 256
  using self in, 254
  using skills in, 254

D

danger
  beginning counselors dealing with, 214
  competence in handling, 216
  consulting with other professionals
    about, 217
  coordination with client's loved ones,
    218
  coordination with client's significant
    others, 218
  in domestic violence situations, 212
  errors of empathy in, 216
  imminent, 32, 154, 187
  interfering with listening, 38
  to others, 206
  preoccupation with liability in, 216
  responding to possible communications
    about, 190
  seriousness of, 214
  unknown, 218
  weight of decisions about, 214
defensiveness, 31
depression, 230
dialectical behavioral therapy, 49
dignity, in managing crises, 186
discrepancies, reflection of, 23
divorce of parents
  and family history, 133
  history of, 131
  reasons for coming to counseling now,
    132
  social resources for, 134
  treatment plans for, 137
  understanding situation, 135
domestic violence
  assessment of, 212
  high-risk behaviors in, 213

  involving children, 213
  physical, 212
  safety plans in, 213
  triggers of, 213
drug use, 133

E

Ellis, Albert, 71
emotions
  as connection to core self, 50
  experiencing, 52, 89
  full expression of, 88
  necessary for growth, 14
  safe environment for, 109
  scariness of exploring, 50
  usefulness of, 14
empathy, 1, 118
  in any relationship, 49
  balancing with UPR, 102–123, 114
  barriers to, 54
  client's joy at, 52
  as core concept, 49
  and counseling-related assessment, 129
  in crises, 215
  deep, 48
  defined, 46, 156
  as distinct from a thought process, 46
  as distinct from sympathy, 47
  errors in, 216
  expressing, 60–77, 289–290
  in filial therapy, 53
  focus activity for, 44, 103
  in furthering communication, 53
  in furthering connection, 53
  genuineness in, 102–123
  getting out of the habit of, 54
  importance of, 50
  intricacies of, 47
  journaling about, 45
  leading to more and deeper emotions, 55
  and letting go of control, 56
  literature regarding, 48
  in managing crises, 187
  with new client, 126
  not equaling agreement, 72
  power of, 50
  reactions interfering with, 111
  and self-awareness, 51
  and self-experience, 51
  straying from, 44
  striving for, 43–59
  in tandem with UPR, 81
  uniqueness of, 52
  when hospitalization is suggested, 201

empathy sandwich, 275
empowerment, resulting from therapeutic
   listening, 38
ending therapeutic relationship
   alternative plans for, 232
   arbitrarily, 234, 238
   beginning counselors' difficulties with,
      243
   with big bang, 241
   client reluctance about, 244
   clients adjusting pace in awareness of, 238
   by clients for unknown reasons, 245
   clients more okay with than counselor,
      245
   counselor self-blame around, 245
   counting down to, 233, 237
   happiness-sadness of, 243
   with initially reluctant client, 227
   last meeting of, 233
   letting clients know they may return,
      233
   with no-show, 241
   not wanting to let go of client, 243
   planning for, 224
   potential for feeling raw in ending
      mid-work, 240
   prematurely, 236
   review for client progress, satisfaction,
      decisions toward, 225
   review for client readiness for, 225
   reviewing client intentions for, 226
   reviewing progress before, 226
   reviewing satisfaction before, 226
   satisfaction and joy in, 246
   seeking feedback in, 242
   suggesting continued work and progress
      following, 237
   telling clients how you *see* them, 233
   wanting too much, 243
environment, interaction with, 8
evaluation
   loss of, 16
   by others, 89
existence, awareness of, 12
existentialism, 49
experience
   as best communicator, 6
   as best teacher, 6
   interpretation of, 10
expressing empathy
   by accepting corrections, 68
   avoiding assessment statements in, 68
   avoiding counselor's hidden agenda in,
      69
   avoiding "me too" statements in, 69

avoiding "must feel" statements in, 69
by body language, 62
common difficulties, pitfalls, and dead
   ends in, 73
do's and don'ts of, 64
by facial expression, 62
by feeling what client feels, 65
focus activity for, 61
focusing attention on client's emotions,
   65
focusing secondarily on client's thoughts
   and actions, 65
having limited vocabulary for feelings,
   74
having personal confidence and faith in
   counseling process, 75
lack of UPR, 75
let client do the talking, 69
by matching client's tone, 62
nuances of, 70
problem with claiming understanding
   or shared experiences, 74
reflect client feelings through your
   natural body language, 66
reflect client feelings with words, 66
reflect implied client feelings and
   underlying thoughts, 66
responding to implied emotions, 70
responding to unpleasant emotions, 71
restating client feelings and underlying
   thoughts, 67
spontaneity in, 70
stating empathy, 67
thinking of the word, rather than feeling
   with client, 73
tone in, 70
trying too hard to get it just right, 73
ways of, 62
when to respond more to content, 72
when to respond more to emotions, 72
with words, 63

F

families
   history of, 133
   using therapeutic relationship skills, 277
feedback, seeking in final meetings, 242
feelings
   avoiding, 54
   fear of, 54
filial glow, 53
filial therapy, 278
   empathy in, 53
five-minute time warnings, 154

G

genuineness, 122
    with clients who are hard to like, 115
    communication of, 110
    concept of, misunderstanding of, 114
    and counseling-related assessment, 129
    declarations of, 110
    defining, 104
    difficulty of, 114
    in empathy, 107
    empathy in, 118
    importance of, 107
    literature supporting and clarifying, 105
    maturity and, 115
    with new client, 126
    observational skills in, 118
    self-development and, 115
    therapeutic listening in, 118
    in UPR, 108
goals
    client's problems with, 145
    communicating to client, 148
    counselor's problems with, 144
    establishing, 145
    reasonable, 146
    unreasonable, 147
Guerney, B. G., 277
guilt, 82

H

heart of counseling, defined, 1–4
Heart of Counseling
    eleven concepts of, 5–18
    how to use, 4
    theoretical base and background of, 3
hospitalization, 201
    counselor responsibility in, 202
    laws, guidelines, and procedures for, 201
    paying for, 202
    reasons for, 201

I

impulse control, lowered, 193, 195
incongruence, 108
independence, in therapeutic relationship, 223
infants, egocentrism of, 11
information gathering, 129
inhibitions, lowered, 193, 195
initial impressions, 139
initial session
    information gathering in, 127

learning about, 126
    planning for ongoing counseling in, 127
    report on, 126
    sample form for, 291
    understanding person or concerns in, 135
initial session report, 126
    academic functioning in, 134
    additional notes for, 139
    family history in, 133
    history of problem in, 131
    identifying information for, 131
    initial impressions in, 135
    listing previous interventions in, 131
    noting alcohol or drug use in, 133
    noting medical concerns in, 133
    presenting concerns in, 131
    presenting problems in, 131
    reasons for coming to counseling at
        present time, 132
    social resources in, 134
    stress in, 134
    treatment plans in, 137
    work functioning in, 134
    writing, 130
insight, overrating of, 37
intake, 126
integrity, in managing crises, 186
intellectual overload, 6
interaction, supportive, 8
interest in others, 84
internal world, mastery of, 16
interventions
    least-restrictive, 186
    in managing crises, 186

J

job task negotiation, 274
Jungian concepts
    of dark side, 72
    of personal shadow, 72

K

kindness, 84
Klein, Armin, 99

L

lack of motivation
    and family history, 133
    history of, 132
    reasons for coming to counseling now, 132

social resources for, 134
treatment plans for, 138
understanding situation, 136
language, culturally laden, 27
liability, preoccupation with, 216
linear thinking, 260
listening
outside of counseling, 36
therapeutic. *See* therapeutic listening
loved ones, counselors consulting with,
273

M

magic pills, 259
maturity, blocking of, 9
medical concerns, 133
messages, 106
modeling, 88, 108

N

National Institute for Relationship
Enhancement, 275
nervousness, 231
new client, 124–149
anxiety of, 141
common dilemmas with, 141
common situations with, 141
confidentiality of, 142
explaining reasons for referral to, 143
focus activity for, 125
helpful explanations for, 142
information for, 142
initial session report on, 126
presenting problem of, helpful information related to, 144
talking about goals with, 144
who needs to know what to expect, 141
nonlistening, 36
non–self-harm agreement, 194
Nordling, Bill, 275

O

observational skills, 118

P

Paley, Vivian, 84
panic, in crisis management, 218
paraphrasing, 22
parents, counselors consulting with, 273
physical violence, 212
prizing, 82, 83

process, trusting, 169
professionals from other fields, counselors
consulting with, 273
progress
continued, 237
definition of, 228
examples of, 229
many forms of, 228
Project Special Friend, 276
public speaking avoidance
and family history, 133
history of, 132
reasons for coming to counseling now,
132
social resources for, 135
treatment plans for, 138
understanding situation, 136

Q

questioning, 9
questioning tone, avoidance of, 29
questions
avoidance of, by therapist, 29
on information previously discussed,
180
of interest to therapist, 40
open, 180
slipping into, 173
quick fixes, 259

R

rational emotive therapy, 114
reasoning, 9
referrals, 143, 179
reflections
during crises, 188
as declarative statements, 28
focus of, 33
higher-level, 162
interfering with listening, 38
levels and nuances of, 22
pedantic, 172
phrasing of, 113
short, 32
of small emotions, 173
sounding like "aha" conclusions, 172
of themes, 176
with time warnings, 154
verbatim, 32
rejection, 11
relationship break-up
and family history, 133
history of, 131

relationship break-up (*continued*)
  reasons for coming to counseling now, 132
  social resources for, 134
  treatment plans for, 137
  understanding situation, 135
Relationship Enhancement Model, 277
religious tenets, 81
responsibility, awareness of, 13
Rogers, Carl, 3, 268
  core conditions of, 108
  on empathy, 48
  on genuineness, 104
  on unconditional positive regard, 79

S

safe environment, 88, 92, 109
safety
  client's ability to guarantee, 194
  helping clients plan for, 189
safety plan, 194
  avoiding elements of suicide plan, 194, 195
  in domestic violence situations, 212, 213
  immediate contact in, 196
  time-specific, 194
  trusting client to follow, 216
  uncertainty about, 215
school counselors, 269, 271, 272, 274, 275
self-acceptance, 87
  counselor's lack of, 95
self-actualization
  blocking of, 9
  concept of, 7
self-awareness, empathy as aid to, 51
self-care, 128
self-concept, development of, 10
self-harm, 194
  non–self-harm agreements, 186
self-immersion, 256
self-knowledge, 256
self-perpetuating cycle, 81
self-responsibility
  anxiety associated with, 13
  maintaining, when hospitalization is suggested, 201
  in managing crises, 186
self-worth, questions of, 12
sexual molestation, 230
shoulds, 71
  of counselors, 116
silences
  allowing clients to own, 35
  client discomfort with, 159

social resources, 134
stress, major areas of, 134
suffering, purpose of, 15
suicide, 189, 190
  asking other professionals for help with, 197
  lethal plan for, 192
  means of, 192
  means of, getting rid of, 195
  mild ideation of, 197
  plan for, 191, 194, 195
  preventive factors for, 192
  previous attempts at, 193
  as reason to break confidentiality, 196
  strong ideation of, 202
  working slowly and carefully to prevent, 196
summarizing, 24
support groups, 128

T

teachers, counselors consulting with, 272
themes, reflection of, 22
therapeutic listening, 118
  accepting corrections in, 34
  allowing clients to own most silences in, 35
  avoidance of questions or questioning tone in, 29
  behaviors involved in, 26
  body language in, 26
  common errors and problems in, 38
  common interfering thoughts in, 38
  and counseling-related assessment, 129
  culturally laden language and, 27
  differences in, from listening outside of counseling, 36
  differences in, from nonlistening, 36
  difficulty of, 38
  do's and don'ts of, 25
  focus activity for, 20
  focusing reflections during, 33
  genuineness of, 107
  how and why it works, 36
  interruptions in, 34
  multitasking with, 41
  with new client, 126
  oddness of, 39
  provoking anxiety during, 37
  reflection of perception of what client communicates, 27
  reflections as declarative statements in, 28
  returning to important communications during, 41

rich and subtle skills of, 19–42
short reflections in, 32
tuning in during, 21, 38
25/75 ratio in, 33
and urge to fix things immediately, 39
verbatim reflections in, 32
while expressing empathy, 61
therapeutic relationship skills
in classrooms, 275
in consultation, 272
developing self through, 268
further study of, 278
growing, 266–279
in job task negotiation, 274
teaching clients to use in their relationships, 277
unexplored potential of, 278
therapeutic relationships
across cultures, 250–265
asking questions in, 178
basing questions on information discussed in, 180
clients' difficulties in starting, 180
difficulty in starting, 167, 168
dispelling need for problem or profundity in, 177
ending, 221–249
explaining to persons significant to
client, 164
focus activity for, 151
forgetting to use, in crisis situation, 218
having clients decide to continue each
week, 232
initial and ongoing structuring of,
150–166
logistics of, 151
in managing crises, 185
open questions in, 180
principle of independence in, 223
recognizing client's use of, 170
reviewing skills for, 173
sharing experience of letting go in, 177
skills for managing clients' crises,
183–220
standard blocks of time for, 232
suggesting common areas of importance
for discussion in, 179
suggesting topics in, 178

time warnings, 154, 155
treatment plans, 137, 139

U

unconditional negative regard, 85
unconditional positive disregard, 85
burnout leading to, 93
unconditional positive regard (UPR)
balancing with empathy, 102–123
barriers to having and communicating,
92
communication of, 90
consistency of, 80
and counseling-related assessment, 129
defining, 82
focus activities for, 79
focus activity for, 103
as full expression of emotions, 88
genuineness in, 108
genuineness of, 107
importance and power of, 86
journaling on, 79
lack of, 75
literature supporting and clarifying, 86
with new client, 126
outside of counseling, 97
outside of counseling relationships, 86
poem about, 99
providing in genuine manner, 102–123
reactions interfering with, 111
rewards for client, 89
rewards for counselor, 89
striving for and communicating, 78–101
in tandem with empathy, 81
understanding, defined, 156

V

videotaping, 44

W

warmth, 82
work
continued, 237
defined, 147
work functioning, 134